AGRICULTURE ISSUES AND POLICIES

Additional books in this series can be found on Nova's website under the Series tab.

Additional E-books in this series can be found on Nova's website under the E-book tab.

AGRICULTURE ISSUES AND POLICIES

U.S. AGRICULTURAL PREPAREDNESS AND THE AGRICULTURE RESEARCH ENTERPRISE

WARREN C. YOUNG
EDITOR

nova publishers
New York

Library of Congress Cataloging-in-Publication Data

ISBN: 978-1-62618-437-4

Published by Nova Science Publishers, Inc. † New York

CONTENTS

PREFACE

The United States is the undisputed world leader in agricultural production today, but as we look out across the 21st century, agriculture faces a series of challenges: managing new pests, pathogens, and invasive plants; increasing the efficiency of water use; reducing the environmental footprint of agriculture; growing food in a changing climate; managing the production of bioenergy; producing safe and nutritious food; and assisting with global food security and maintaining abundant yields. Meeting these challenges requires a new commitment to research, innovation, and technology development in agriculture. This book examines the challenges of agricultural preparedness and the required public commitment to agricultural research, fostering a culture of innovation and excellence to address some of the greatest threats to U.S. long-term prosperity and security.

Chapter 1 – In January 2011, Catherine Woteki, the Undersecretary for Research, Education, and Economics, and Roger Beachy, thenDirector of the National Institute for Food and Agriculture, spoke to the President's Council of Advisors on Science and Technology (PCAST) and subsequently requested an examination of areas of critical need within agriculture. In March 2012, the Secretary of Agriculture addressed PCAST and noted the critical role of the U.S. agricultural research enterprise was to prepare the Nation to meet future challenges. He asked PCAST to identify and recommend ways to optimize innovative public sector agricultural research, to stimulate training of the workforce, and to ensure that the United States maintains its leadership in agriculture around the world. In response, PCAST formed a working group of PCAST members and experts from the public and private sectors to address the charge and to formulate recommendations. Informed by the working group's deliberations, PCAST has developed this report to provide advice to the Administration about Federal actions that can promote agricultural research innovation.

Our most important conclusion is that our Nation's agricultural research enterprise is not prepared to meet the challenges that U.S. agriculture faces in the 21st century for two major reasons. First, PCAST finds that the proportion of Federal funding for agricultural research allocated through competitive mechanisms is far below the proportion in other agencies, which fails to adequately encourage innovation. Second, PCAST finds that the current agricultural research portfolio is not optimally balanced; it overlaps with private sector activities in several significant areas, while underfunding other important areas that are not addressed through private efforts.

Chapter 2 – The U.S. Department of Agriculture (USDA) Research, Education, and Economics (REE) mission area has the primary federal responsibility of advancing scientific knowledge for agriculture through research, education, and extension. USDA REE responsibilities are carried out by four agencies: the Agricultural Research Service (ARS), the National Institute of Food and Agriculture (NIFA), the Economic Research Service (ERS), and the National Agricultural Statistics Service (NASS). The USDA administers extramural federal appropriations to states and local partners primarily through three funding mechanisms: formula funds, competitive grants, and non-competitive grant programs.

The FY2012 Agriculture Appropriations Act (P.L. 112-55, H.R. 2112) contained $2.53 billion in discretionary funds for USDA agricultural research, education, and extension programs. While inflation-adjusted public spending for agricultural research grew steadily from the 1950s to the 1970s, it has remained relatively flat since the 1970s (with a few exceptions), and growth in funding has lagged behind that of other national science agencies.

The enacted 2008 farm bill (P.L. 110-246) directed USDA to reorganize the REE mission area. The farm bill created a new entity called the National Institute for Food and Agriculture (NIFA), which assumed all programs and authorities from the Cooperative State Research, Education, and Extension Service (CSREES). A new competitive grants program for basic and applied research, called the Agriculture and Food Research Initiative (AFRI), was also established by the 2008 farm bill and is administered by NIFA. The farm bill also extends and expands mandatory and discretionary funding for specialty and organic crops research, bioenergy programs, and pollinator protection programs, among others. Several mandatory programs (research and otherwise) that were authorized in the 2008 farm bill do not have a budget baseline that extends beyond the end of the 2008 farm bill (September 30,

2012). If policymakers want to continue these programs in the next farm bill, they will need to pay for them with other offsets.

Debates over the direction of public agricultural research and the nature of its funding mechanism continue. Ongoing issues include the need, if any, for new federal funding to support agricultural research, education, and extension activities, and the implications of allocating federal funds via formula funds versus competitive grants. In addition, factors including the growing importance of specialty crops, international trade negotiations, and a renewed interest in international agricultural development have many groups believing that Congress needs to increase support of U.S. agriculture through expanded research, education, and extension programs, whereas others believe that the private sector, not taxpayer dollars, should be used to support these activities.

While the 2008 farm bill provided a significant funding boost for agricultural research, relatively speaking, it is unclear whether budgetary resources and political will can sustain funding for agricultural research and related activities.

Chapter 3 – This brief examines the funding and performance of agricultural research and development (R&D) to assess the evolving roles of the public and private sectors in the U.S. agricultural research system. There is a clear long-term trend toward greater private-sector funding and performance of R&D. In 2007, the private sector performed 53 percent of total food and agricultural research in the United States, and privately funded R&D has grown faster than publicly funded R&D over the long term

Chapter 4 – Major Findings:

- By 2050, global agricultural demand is projected to grow by 70-100 percent due to population growth, energy demands, and higher incomes in developing countries. Meeting this demand from existing agricultural resources will require raising global agricultural total factor productivity (TFP) by a similar level. Maintaining the U.S. contribution to global food supply would also require a similar rise in U.S. agricultural TFP.
- TFP growth in U.S. agriculture is predicated on long-term investments in public agricultural research and development (R&D). Productivity growth also springs from agricultural extension, farmer education, rural infrastructure, private agricultural R&D, and technology transfers, but the force of these factors is compounded by public agricultural research.
- The rate of TFP growth (and therefore output growth) of U.S. agriculture has averaged about 1.5 percent annually over the past 50

years. Stagnant (inflation-adjusted) funding for public agricultural research since the 1980s may be causing agricultural TFP growth to slow down, although statistical analyses of productivity growth trends are inconclusive.

- ERS simulations indicate that if U.S. public agricultural R&D spending remains constant (in nominal terms) until 2050, the annual rate of agricultural TFP growth will fall to under 0.75 percent and U.S. agricultural output will increase by only 40 percent by 2050. Under this scenario, raising output beyond this level would require bringing more land, labor, capital, materials, and other resources into production.
- Additional public agricultural R&D spending would raise U.S. agricultural productivity and output growth. Raising R&D spending by 3.73 percent annually (offsetting the historical rate of inflation in research costs) would increase U.S. agricultural output by 73 percent by 2050. Raising R&D spending by 4.73 percent per year (1-percent annual growth in inflation-adjusted spending) would increase output by 83 percent by 2050.

Chapter 5 – Over the last several decades, the U.S. agricultural sector has sustained impressive productivity growth. The Nation's agricultural research system, including Federal-State public research as well as private-sector research, has been a key driver of this growth. Economic analysis finds strong and consistent evidence that investment in agricultural research has yielded high returns per dollar spent.

These returns include benefits not only to the farm sector but also to the food industry and consumers in the form of more abundant commodities at lower prices.

While studies using different methods and coverage give a range of estimates of returns to agricultural research, there is a consensus that the payoff from the government's investment in agricultural research has been high.

In: U.S. Agricultural Preparedness ... ISBN: 978-1-62618-437-4
Editor: Warren C. Young

Chapter 1

REPORT TO THE PRESIDENT ON AGRICULTURAL PREPAREDNESS AND THE AGRICULTURE RESEARCH ENTERPRISE*

Executive Office of the President, President's Council of Advisors on Science and Technology

EXECUTIVE SUMMARY

The United States is the undisputed world leader in agricultural production today, but as we look out across the 21st century, agriculture faces a series of challenges:

- Managing new pests, pathogens, and invasive plants.
- Increasing the efficiency of water use.
- Reducing the environmental footprint of agriculture.
- Growing food in a changing climate.
- Managing the production of bioenergy.
- Producing safe and nutritious food.
- Assisting with global food security and maintaining abundant yields.

---

* This is an edited, reformatted and augmented version of the President's Council of Advisors on Science and Technology, dated December 2012.

Meeting these challenges requires a renewed commitment to research, innovation, and technology development in agriculture. Private industry will continue to play an important role meeting these challenges in areas directly related to commercial developments and commodities. But many of the developments necessary to meet these challenges are public goods and not easily monetized. These challenges require a strong public commitment to agricultural research, one that fosters a culture of innovation and excellence to address some of the greatest threats to U.S. long-term prosperity and security.

In January 2011, Catherine Woteki, the Undersecretary for Research, Education, and Economics, and Roger Beachy, thenDirector of the National Institute for Food and Agriculture, spoke to the President's Council of Advisors on Science and Technology (PCAST) and subsequently requested an examination of areas of critical need within agriculture. In March 2012, the Secretary of Agriculture addressed PCAST and noted the critical role of the U.S. agricultural research enterprise was to prepare the Nation to meet future challenges. He asked PCAST to identify and recommend ways to optimize innovative public sector agricultural research, to stimulate training of the workforce, and to ensure that the United States maintains its leadership in agriculture around the world. In response, PCAST formed a working group of PCAST members and experts from the public and private sectors to address the charge and to formulate recommendations. Informed by the working group's deliberations, PCAST has developed this report to provide advice to the Administration about Federal actions that can promote agricultural research innovation.

Our most important conclusion is that our Nation's agricultural research enterprise is not prepared to meet the challenges that U.S. agriculture faces in the 21st century for two major reasons. First, PCAST finds that the proportion of Federal funding for agricultural research allocated through competitive mechanisms is far below the proportion in other agencies, which fails to adequately encourage innovation. Second, PCAST finds that the current agricultural research portfolio is not optimally balanced; it overlaps with private sector activities in several significant areas, while underfunding other important areas that are not addressed through private efforts.

PCAST recommends the creation of a new innovation ecosystem for agriculture that leverages the best from different parts of the broad U.S. science and technology enterprise, focusing public investment on challenges that enhance the public good and are not readily done by the private sector alone. PCAST calls for a strategic investment that will

create the path toward an improved innovation ecosystem for the U.S. heartland, enhancing the economy and harnessing the power of U.S. innovation in science and technology to address this set of great challenges to the Nation.

A new agricultural research enterprise should be centered on competitive intramural and extramural research efforts that bring together scientists from traditional agricultural fields and those from the biological and physical sciences. A focused public investment would not only invigorate agricultural research and create opportunities for new business ventures funded by the private sector, but also provide the means to train the next generation of farmers and agricultural researchers to meet the workforce demands of U.S. agriculture in the 21st century.

Creating an innovation ecosystem for agricultural research requires investment in three main areas: research support, training and workforce development, and research infrastructure. In all three areas, PCAST has considered how to enhance and invigorate the existing structure while preserving the strengths and important roles of existing programs. In some cases, these reforms will require new funding to achieve the stated goals. In other cases, a restructuring of existing funds can enhance the use of competition for distribution of funding and can expand the Federal research portfolio with respect to issues of public good.

Overall, PCAST recommends that the United States increase its investment in agricultural research by a total of $700 million per year as detailed below, focusing on addressing the emerging chal-lenges described above.[1]

Overview of PCAST Recommendations to Improve Agricultural Preparedness and Maintain U.S. Leadership in Agriculture

1. Expand the role of competition in agricultural research funding:
    a. Expand the use of competition in allocation of research funding within intramural and extramural programs of the U.S. Department of Agriculture (USDA).
    b. Increase the National Science Foundation (NSF) budget for basic science relevant to agriculture from $120 million to $250 million per year.

c. Increase the USDA budget for competitive funding of extramural research from $265 million to $500 million per year, consistent with the 2008 congressional authorization.

2. Greatly expand a competitively awarded fellowship program for graduate students and postdoctoral researchers at a level of $180 million per year with at least 5year funding.
3. Expand the USDA program of competitive awards for new infrastructure investments for agricultural research with an emphasis on specialization and consolidation to avoid redundancies.
4. Create six large, multidisciplinary innovation institutes focused on emerging challenges to agriculture, supported by publicprivate partnerships at an initial new Federal investment of $150 million per year to create six institutes at a funding level of $25 million per year for no less than 5 years.
5. Conduct an internal review of Federal regulatory policy for agriculture to promote regulatory clarity, consistent with Executive Order 13563, as well as the Presidential Memorandum on technology transfer from the national laboratories to the marketplace.
6. Establish an implementation committee to act on these recommendations. Create a permanent, independent science advisory committee to advise the Chief Scientist of the USDA.

I. AGRICULTURAL RESEARCH AT A CROSSROADS

Since its earliest colonial history, agriculture has played a central role in the social and economic activity of the United States. Since that time, the Nation has depended on agriculture not only to feed its citizens, but also as a major driver of its economy. Exports of agricultural products produced a $34 billion trade surplus in 2010 and a $37 billion trade surplus in 2011,[2] and the agricultural sector is currently responsible for 1 in 12 American jobs.[3] Beyond its economic impact, U.S. agriculture provides a foundation for world food stability and security, supplying most of the food aid to developing nations around the world. Looking to the future, U.S. agriculture must continue to be the backbone for the emerging U.S. bioeconomy, helping the Nation meet its need for sustainable sources of energy and materials, and simultaneously contributing to the prosperity of rural communities. A vibrant U.S. agriculture enterprise is paramount to the future wellbeing of the Nation.[4]

U.S. prominence across many different areas of agriculture derives in part from a rich history of commitment to agricultural research. Our current agricultural research system dates to 1862, when President Lincoln signed into law two pieces of legislation creating the USDA and the network of Morrill Land Grant Colleges. Twentyfive years later, in 1887, the role of agricultural research in supporting and promoting the rural economy was established with the passage of the Hatch Experiment Station Act, which jointly established agricultural research outposts and expanded land grant universities. Then in 1917, the United States recognized that the fruits of agricultural research must be applied to the field and shared with farmers in the Cooperative Extension Act.[5]

Research at the landgrant colleges was organized around local problems and local knowledge. This is partly due to the parochial nature of agriculture, which requires knowledge of biological interactions within the framework of local weather, soil, and pests. At the same time, the wide distribution of agriculture research across the country also served to translate new research into practice, as agricultural institutions, including the USDA Cooperative Extension Service, were designed to supply research findings to farmers in the form of local management practices for plant and animal varieties, feed, fertilizers, and pest management.

Agricultural institutions also led the Nation toward the democratization of education. For example, George Washington Carver received his education as the first African American college student (and later, the first African American faculty member) at Iowa State University. The opportunities he was afforded were not unique, but shared by a multitude of students from working-class communities across the country who received levels of education previously unthinkable in rural areas.

Agricultural research has helped to make the U.S. farmer among the most efficient in the world. Today, the United States stands as the global leader in meeting the world's demand for food, thanks to significant productivity gains achieved since the middle of the 20th century. The American agricultural enterprise has consistently boosted productivity over the past few decades for most major crops and livestock, and it has been a hallmark of industrial innovation. Since the 1930s, yields of staple crops such as soybeans have more than tripled despite a decrease in available cropland area, and corn yields have increased fivefold since the 1930s.[6] Numerous studies attribute these successes to the discovery, development, and rapid adoption of new technologies and agronomic practices, which directly stem from the U.S. investment in basic and applied plant, animal, and agricultural research across the Federal agencies and its translation to farms by both the land grant universities and the private sector.[7,8]

Public financial support for agricultural research has waned over the past three decades (relative to the increases of the 1960s and 1970s) as other areas of science and technology research and development (R&D) have seen substantial growth. Public funding of agricultural research, in real dollars, has remained at nearly the same level for the last 30 years.[9] (Note that other Federal agencies, particularly the National Science Foundation (NSF) and the Department of Energy (DOE), have provided additional support for basic research underpinning the explicitly agricultural mission of USDA). Excluding recent research on biofuels production, less than $500 million per year is available for competitive grants in agriculture, roughly 2 percent of the amount of competitive funding from the National Institutes of Health and 6 percent from the NSF (see Chapter 3 for more detail).[10,11] One consequence of the small amount of competitive funding for agricultural research is the decline in the training of new agricultural scientists and the hindered recruitment of a new generation of the best young scientists into this area. Indeed, recruitment of students into agricultural research is one of the major challenges to the U.S. agricultural research enterprise in the coming decades.

The success of the agricultural industry over this period, despite the paucity of public investment and declining workforce, can be seen in part as a residual benefit of the long history of agricultural research investment, but it also reflects a significant shift from public to private dominance in agricultural research and development. Indeed, private industry now outspends the USDA in agricultural research by more than three to one (see Chapter 3 and Figures 1 and 2).

With this shift toward the private sector has come a narrower focus—the vast majority of agricultural research funding is now spent on a very small number of major or commodity crops, in particular corn and soybeans. The shift from public to private investment in R&D continues to drive remarkable innovation, especially in those areas that are most directly related to product development over a relatively short time frame consistent with industry's profit timeframe. For example, major seed companies have thousands of Ph.D. and M.S.trained scientists working in plant breeding to develop new strains of corn and other major crops. This workforce includes plant breeders, geneticists, field technicians, and laboratory personnel supporting activities such as genotyping. In light of this private investment, there are significant questions about the role and nature of public research investment in agricultural research. Specifically, it is not clear that the current public investment is being optimally used, especially where some public investments overlap research topics supported by industry.

Although the United States is the undisputed world leader in agricultural production today, continued innovation and investment are essential to maintaining a competitive advantage in the future. The private sector's commitment to agricultural research in the United States remains strong. However, many of the most important companies for agricultural research are large international corporations; many of them are investing and even outsourcing significant research dollars overseas, as China, India, and Brazil start to make large public investments in agricultural research.[12] The waning public investment in agricultural research in the United States contributes significantly to the risk of losing its international leadership in agriculture.

As we look across the 21st century, we see that agriculture faces a series of new challenges that will require a renewed commitment to innovation and advanced technology development. Private industry will play an important role in the research required to meet these challenges, just as it does today in areas directly related to product development. But much of the necessary research is unlikely to result in new products in a time horizon short enough to incent the private sector to shoulder the entire research burden. Moreover, many of these challenges are clearly in the public domain, as they focus on critical public goods, such as long-term water security; integrated pest-management strategies; or the development of new varieties of livestock, cereal, vegetable, and cover crops that commercial enterprises may not have an interest in. In many cases, important benefits of agricultural research cannot be monetized, making them an unlikely focus for the private sector.

These challenges require a public commitment to reimagining agricultural research in this country. The United States requires an agriculture system that (1) meets its population's needs in long-term balance with its natural resource base and (2) fosters a culture of innovation and excellence to address some of the greatest threats to its long-term prosperity and domestic and international security. In this report we explore these emerging challenges to agriculture and examine the roles that public and private investments in agricultural research can contribute to the sustainability of agriculture over the next century. We also examine the structure of the existing Federally supported agricultural research enterprise and discuss possible ways to reshape public investment in agricultural research to best face the challenges ahead.

We note that agricultural research is only one of the ways that the Federal Government influences agricultural activities in the United States. A wide variety of policies, including farm subsidies and regulatory policy, are also important components of an innovation ecosystem for agriculture that we describe in this report. For simplicity, we do not discuss these components of agricultural policy in this report, but we recognize that they, like the agricultural research enterprise discussed below, must be subject to review and revision to encourage the greatest benefits from agricultural research to flow from the laboratory to the farm.

As we worked on this report, the Nation's farmers are suffering through one of the worst droughts in U.S. history. Low snowfall last winter, record high temperatures last spring, and heat waves and drought this summer have combined to decimate the corn and soybean crops across the Midwest. Corn prices are at record high levels, with over 2,000 U.S. counties in 31 states designated as disaster areas by the USDA. Agricultural research cannot immediately protect the farmer whose corn is withering in the field, but it may provide future long-term strategies for dealing with heat waves and other extreme weather related problems that are anticipated as climate change proceeds. Further, the need to deal with these growing challenges in agriculture, including new pests and pathogens, controlling agriculture's environmental impact, health and nutritional concerns, and international food security underscores the importance of agricultural research to the health, prosperity, and security of the Nation. Agricultural research that addresses these concerns requires an upfront national investment and commitment to long-term returns on the investment shared by all.

II. New Challenges Require a New Investment in Agricultural Research

Over the next few decades, U.S. agriculture will confront more competition from overseas; struggle with global resource challenges, including limits to the inputs of water, soil, and nutrients; combat new and more virulent pests and pathogens; and be asked to meet the ever increasing demand for improved dietary health, food security, and other agricultural products and materials including biofuels, all in the face of a growing population and changing climate. Successfully overcoming these challenges requires a U.S. agricultural research enterprise that harnesses the newest advances from across the physical and life sciences, builds on a broader public investment in science and technology, and then applies these discoveries to the specific challenges of agriculture.

The major focus of the U.S. agricultural research enterprise over the past two decades in both the public and private sectors has been on increasing yields of major grains and livestock. Crop yields will continue to be a focus for the private sector, but the case for public investment in this area of research and development is less compelling. Although there is certainly a need for more basic research on plant and animal science that may someday contribute to new products and new agronomic practices, it is not clear that Federally-funded research focused solely on increased corn yields, for example, is adding substantially to what the private sector is already doing very effectively. Public investment in agricultural research could shift its attention toward problems that the private sector does not address. In some cases, there are market failures that prevent private investment, even when commercial applications exist; in other cases, the argument for public investment in research and development comes from concerns that are squarely in the public domain.

The emerging challenges to agriculture provide a strong case for a new public investment in agricultural research. These challenges require a change in focus for our Federally funded research programs towards an emphasis on basic research as well as managing the risks associated with emerging threats such as new pests and pathogens, limited water availability, environmental impacts of agriculture on human and environmental health, or adaptation to a changing climate. In the following section we describe some of the challenges that we see as most pressing, along with some of the specific research areas that are worthy of increased public investment. Some of these areas can best be addressed through public private partnerships, where there are strong commercial interests with direct applications to product development. Other areas are entirely in the public domain and just as deserving of vigorous public support because the research outcomes could lead to a more resilient, more sustainable agricultural system, with benefits shared by all.

Managing New Pests, Pathogens, and Invasive Plants

Agriculture is in a constant race to develop new and better strategies for dealing with short-generation pests and disease species that continually evolve resistance to the current control strategies and to changing growing conditions. Whether focused on major grains, specialty or cover crops, or livestock, these pests, parasites, and pathogens represent a major risk to successful food production. For crops, weeds that have developed resistance to conventional herbicides are a serious threat to sustaining high yields. For animals, bio surveillance and food safety are critical concerns for U.S. agriculture and food systems. And for both crops and livestock in the 21st century, new insights regarding the roles of beneficial microbial communities may stand to boost health and productivity.

A good example of the threat of new plant pathogens is the problem of wheat stem rust, a fungal disease (*Puccinia graminins* f.sp. *tritici*) that can reduce normal wheat yields by 70 percent. Wheat varieties resistant to these diseases were bred in Mexico in the 1940s and 1950s and distributed over millions of hectares in the 1960s and 1970s, thus starting the "green revolution." But following the success of this program, resistance breeding for stem rust ceased to be a priority. Today, a new stem rust, Ug99 (a stem rust race identified from Uganda in 1999) poses a worldwide threat to wheat as this virulent strain begins to move by wind currents out of Africa and into the Middle East. Over 80 percent or more of the worldwide wheat crop is considered susceptible to this rust.[13] The narrow genetic base for wheat in the United States may severely compromise food supplies and stability if this disease gains a foothold in North America. Equally important is the potential for disruption of wheat commodity markets worldwide, for example, if the major wheat growing regions of the Punjab are infected by Ug99. And wheat stem rust is only one example of a serious new disease threatening a major crop. The result of such a catastrophe is the potential for an exacerbated security situation in the Middle East, South Asia, and other restive parts of the world.

Specialty crops are also put at risk by new plant pathogens that have no effective management strategy. For example, citrus greening disease, caused by a bacterium (*Candidatus Liberibacter spp.*) and spread by an insect, was detected in Florida for the first time in 2005 and now threatens the state's citrus industry, putting at risk the $9.3 billion in economic benefits it provides. The disease was detected in California and Texas earlier this year[14] and endangers western citrus production as well.[15] Another example is Pierce's disease of grapes, caused by *Xylella fastidiosa,* which endangered the wine industry in California after its vector, the invasive glassy winged sharpshooter, arrived in 1996.[16] In both of these cases, there is no detailed knowledge of how the bacterial pathogen causes disease or how plants might resist it. Furthermore, insecticide treatments can result in insecticide resistant insects, decreased numbers of beneficial insects, and groundwater contamination. As with wheat stem rust, these two diseases have already caused disruptions of global food and wine production, leading to economic consequences with widespread ripple effects. Protecting crops of all kinds against emerging diseases requires an understanding of the basic functional principles of the plant immune system and its deployment in a variety of contexts, as well as the ecological relationships between plants, pathogens, and other organisms, and their interactions with the local physical environment.

New, emerging, and evolving pathogens are also a challenge to animal health and production. Although improved management and better diagnostic and prophylactic tools are now available, emerging diseases and changes in animal management will always create new opportunities for old diseases. Many of the best therapeutic agents heavily used in controlling the more problematic pathogens are losing their effectiveness due to evolved resistance by the pathogens. For example, the pathogenic roundworm of ruminants (which include production livestock such as cattle and sheep), *Haemonchus contortus*, has developed resistance to every available drug. Public demands for the humane treatment of animals and against the nonmedical use of antibiotics also make it difficult to fight many diseases in conventional ways, potentially contributing to public health challenges as a result of antimicrobial resistance andthe spread of deadly antibiotic-resistant pathogens.

Using a range of discoveries in basic molecular biology and genetics, new approaches must be developed to deal with the problem of resistance to treatments of both plant and animal diseases. In addition, new treatment strategies must consider the impact of medicines or chemical treatments on the nutrition and health of the consumers and of the environment. This is illustrated aptly by the yearly "essential use allowance" afforded to the ozone-depleting chemical methyl bromide, which is the only effective treatment of soil to protect plants against a variety of fungal and oomycete pathogens.[17] Included here are the increasingly problematic agents of the potato late blight, *Pytophthora infestans*, and other species in this genus that infect oaks, tomatoes, and a variety of important crops. A diversity of safe and effective treatments for a wide suite of pests and pathogens must be developed to ensure that agricultural practitioners have an arsenal of defenses in reserve to protect their crops.

Increasing the Efficiency of Water Use

A second major challenge for agriculture is the reduction of water use. Today, agriculture accounts for 80 percent of the Nation's overall consumptive water use, and in many Western States, it accounts for over 90 percent.[18] In the

Great Plains, recent droughts have substantially depleted the Ogalala aquifer, which runs from South Dakota to Texas, and create the dual problem of high soil salinity and water shortages, thus making the water unavailable or unusable for farmers. In naturally arid lands (particularly the western part of the country), the productivity of irrigated land is approximately two to four times greater than that of rainfed land.[19] But in some of these regions, such as the Southwest where the population is growing rapidly, water is being diverted from agricultural use to meet the water needs of urban communities. In general, as cities draw on more water resources for their rapidly growing populations, agriculture production must significantly improve its water use efficiency, perhaps by shifting crops and changing current agricultural practices.

Reducing agricultural water use will not be easy. For some crops, significant improvements in efficiency are possible through better design of irrigation systems (such as microirrigation, a technique pioneered by the 2012 World Food Prize winner, Daniel Hillel) and the continued development and use of advanced crop management tools. Improvements on a scale that will meet future demands for agricultural production will only come with the development of new crop varieties that are more drought tolerant. Development of these new varieties will result predominantly from major advances in our understanding of basic plant physiology and genetics, particularly as it relates to water use efficiency and photosynthesis.

Research to reduce water use for food production, whether from cropping shifts, improved water management in livestock operations, new plant varieties, or better irrigation technologies, requires a blend of public and private research investment, similar to that required to address the general problem of increasing food production. Major breakthroughs in this area are likely to require a broad research program with investments including germplasm, ongoing monitoring of water resources, weather and climate, and plant and animal physiology rather than targeted, applied research. Such investments are difficult to recover by a single company and yet are essential for meeting a critical national need for reduced vulnerability to drought conditions.

Reducing the Environmental Footprint of Agriculture

A third major challenge for agriculture is to increase production of food, fiber, and fuel while simultaneously decreasing the environmental footprint with respect to fertilizers, pesticides, soil erosion and depletion, pollution associated with livestock production, and agriculture's contribution to greenhouse gas emissions. Modern methods of agriculture, for both crops and livestock, have come at a cost to the environment. For example, fertilizers applied to farms across the middle of the country add to the nitrogen load of waterways in the Midwest, which in turn flow into the Mississippi river and create a dead zone in the Gulf of Mexico,[20] a problem further exacerbated by animal waste disposal.[21] As commodity prices increase and agriculture expands to meet growing food demand, these impacts are likely to grow, with concomitant increased negative effects on ecosystems.[22]

Since the publication of Rachel Carson's *Silent Spring* 50 years ago this fall, concerns about the impacts of pesticides on the health of humans and wildlife have motivated broad research discoveries in the area of integrated pest management and host plant resistance, as well as new product development. Ensuring that chemicals used in agriculture are safe for humans and for the environment is a critical role for government. It requires continued investment in agricultural research focused on the full life cycle of pesticide usage.

Another important impact of the conversion of land from prairie or forest to agriculture is the effect on biodiversity. To keep up with the demand for food from a growing U.S. population and more recently for biofuels, agriculture has converted the native U.S. landscape into pastures and fields. This conversion now covers 46 percent of the U.S. land base.[23] The loss of native landscapes[24] has severely affected native biodiversity. This biodiversity is responsible for essential ecosystem services such as water conservation and flow management, soil conservation, and pollination services essential for agriculture, as documented in our recent PCAST report on sustaining environmental capital.[25]

A further environmental impact of agriculture involves the soil that supports all future plant production. Agricultural practices such as tilling have sometimes reduced soil quality, resulting in soil erosion, carbon and mineral depletion, reduced water infiltration, and increased reliance on chemical fertilizers.[26] Soil erosion may also be affected by climate change that causes more extreme storms and intense rainfalls. A major research challenge is to understand how different agricultural practices affect soil quality and nutrient-enriching microbial communities, identifying tradeoffs, new treatments, and best practices that can enhance soil fertility, making our agricultural system more resilient to a wide variety of stresses.

Finally, agriculture accounts for between seven and eight percent of all U.S. greenhouse-gas emissions, even before accounting for land-use change and downstream emissions associated with food processing and food waste.[27] Some of these emissions are related to the use of nitrogen fertilizers and pollution from animal waste, which can lead to nitrous oxide formation and emission; others are related to rice cultivation and animal production, which lead to methane emissions. Finding ways to reduce these agriculture emissions is an important component of a responsible climate-mitigation strategy. Reducing greenhouse-gas emissions can be achieved through various techniques, including precision agriculture (the highly targeted use of fertilizer), animal dietary manipulation, improved plant and animal productivity, and manure management. But new approaches and practices are needed, and their development will again rest on research.

The big-picture challenge, then, is to develop new management practices that reduce the different environmental impacts attributable to agriculture and to improve and restore the natural resource base such as soil and water, while maintaining a high level of productivity. Some tradeoffs will be inevitable, but new technologies can be used to reduce the environmental impacts of fertilizers, livestock waste, and other inputs by enhancing efficiency and management. New practices are essential for the stewardship of U.S. agricultural lands to ensure that they continue to provide the yields necessary to feed the population, contribute to the U.S. economy, and support the well-being of rural communities. Such research efforts are clearly in the public interest, and the resulting benefits may accrue to the farmer (e.g., through reduced costs), but the greater benefits are an improved environment and improved public health.

Growing Food in a Changing Climate

A fourth major challenge for U.S. agriculture results from changes in temperature and precipitation associated with global climate change. The United States must develop greater resilience to a changing climate through a broad research program aimed at new agricultural strategies to adapt to shifts in weather and climate.

It is likely that no one is more attuned to weather and climate than a farmer. Agriculture depends on the weather in a more direct manner than nearly any other sector of our economy. In fact, U.S. farmers have already witnessed a remarkable change in climate over the past four decades, as planting times have moved earlier and earlier, and even greater changes are predicted for the future. Natural variability is a prominent feature of America's climate, across all of its diverse regions. The American farmer must already prepare for adverse weather conditions, from flooding and late snowfalls in some regions, to heat waves and droughts in others. But climate change, driven primarily by the addition of greenhouse gases to the atmosphere, is changing the nature of the challenge that farmers now face.

One major issue is the gradually warming summer temperatures, and their impact on crop yields. Numerous studies have shown that crop yields can drop precipitously when summer temperatures rise above a critical level—typically 29 to 32 degrees Celsius.[28,29] Such temperatures will become more and more common as Earth warms throughout this century. This effect is illustrated by the European heat wave of 2003, which resulted in a 30 percent decline in regional harvest; the more extreme heat wave in Russia in 2010, which resulted in a 33 percent loss of Russia's wheat harvest;[30] and the U.S. drought of summer 2012, the effects of which are not yet fully known but promise to be dramatic in some regions. As climate continues to change, such heat waves are expected to increase as a result of the shift in average conditions. This presents a formidable challenge to the agricultural system. Not only do our crops need to be bred for regional adaptation to the average growing season, new varieties must be developed that are resilient and tolerant of extremes, particularly during critical periods of plant development. Furthermore, we need to develop new types of interventions to prevent massive losses when such weather events do inevitably occur, such as the Texas drought of 2011 or the Midwestwide drought of 2012.

Climate change is also changing the life cycle and range of pests and pathogens, creating new threats to plant and animal agricultural production. Among these recent threats are the pine blister beetle that is ravaging U.S. forests[31] and soybean rust that recently arrived in the United States from Brazil. Losses due to insects and other pathogens may increase in the coming decades because insect, fungal, and bacterial metabolism should increase in a warmer climate.[32] New research into the basic genetic mechanisms of plant and animal defense systems and subsequent tolerance, resistance, or avoidance strategies for pests and pathogens is desperately needed.

Climate change is also affecting the impact of weeds on agricultural productivity. In some cases, the early emergence of some weeds (due to warmer conditions) results in a dramatic increase in their competitive ability. These weeds might not represent a major threat to crops if they emerge late in the growing season, but earlier emergence can be highly disruptive. Climate change will also alter growing environments, so that they may be more hospitable to invasive plant species such as kudzu, which is currently spreading northward.

Climate change will also directly affect water availability through changes in the patterns and amounts of precipitation. Warmer summer temperatures will also reduce soil moisture, which lessens evaporative cooling, thus amplifying the impact of heat waves. But an even more serious impact of a changing climate on water resources may be the timing of mountain snow melt across the western regions of the United States. As the climate warms, snowpack is expected to decrease, and melting of snowpack will occur over a briefer time interval.[33] This means that spring snow melt will be more intense, driving regional flooding, with reduced stream flow available for agriculture in the summer and fall. Such changes make the challenge of finding more water-efficient plant varieties and developing better irrigation technologies even more important. Early snow melt also extends the fire season,[34] particularly in the Southwest, and it is a contributing factor to the massive fires this year in Colorado and New Mexico.

A silver lining in the impending threat of climate change to agricultural systems is that U.S. farmers have already demonstrated a remarkable ability to adapt to changing weather patterns through a variety of practices, including altered planting times, new crop varieties, use of cover crops, and double-cropping where conditions permit. Research on improving adaptive strategies for a changing climate can lead to greater preparedness for unusual weather conditions. Building resilience of plant and animal production systems to climate change and variability also requires a public investment in extension and education.

Managing the Production of Biofuels and Bioenergy

A fifth challenge for agriculture involves managing a large sector of the bioeconomy, the biofuels industry. The production of fuel from agricultural products has persisted for many years, growing steadily since the 1980s, but the last decade has seen a huge expansion in domestic biofuel production, primarily as corn ethanol. In 2011, the United States produced nearly 14 billion gallons of ethanol, more than a sixfold increase from 2002.[35]

There is widespread concern that the use of arable land for biofuels production competes with food production. To address this concern, there has been a substantial public investment in biofuels research both by the USDA and by the DOE in the production of biofuels from cellulosic feedstocks such as corn stover, components of municipal waste, forest residues, and high biomass crops grown specifically for energy production (in other words, energy crops). Using cellulosic materials does not necessarily solve the problem because some types of biofuels might still compete for arable land. But a shift away from corn ethanol may allow for biofuel production from marginal lands that are not suitable for crop production, and it is likely to increase biofuel yields per acre of land.[36] A continuing challenge is to accelerate basic research on plant physiology and biochemistry, along with making investments in conversion methods for fuel production. This research and investment is needed to enable breakthroughs in biofuel yields. These efforts are largely underway, funded by USDA, DOE, and the private sector, but a series of accompanying challenges has received less attention.

There is a pressing need to expand the scope of agricultural research to include annual and perennial plant species (including trees) that can thrive on marginal land but that have not previously attracted significant interest in the context of biomass production. Managing the growth of the biofuels industry requires a new look at agronomic and forestry practices, such as using water and land resources to balance short-term needs with longer-term questions of sustainability. For instance, in contrast to the large-acreage food crops, most prospective energy crops are perennials that require low inputs and provide improved ecosystem services but entail other challenges such as management of pest and pathogens. In addition, few studies have looked at security issues associated with an expanded bioenergy system, such as vulnerabilities associated with droughts or other extreme weather events. The expanded use of bioenergy is an enormous opportunity for U.S. agriculture and other sectors of the bioeconomy, but supporting food and energy needs will require continued research not only on biofuel production but also on how our agricultural system and energy infrastructure can best be structured to incorporate new markets and new practices.

Safe and Nutritious Food

A sixth major challenge for the U.S. agricultural enterprise is to continue to provide high-quality and safe food to enhance nutrition in the face of growing levels of obesity and diabetes and the emergence of new types of food-borne diseases. The U.S. food system is among the safest in the world. When an outbreak of a food-borne illness does occur, it receives widespread attention, in part because the U.S. public is so used to having nearly universal access to safe food products. The USDA and the Department of Health and Human Services (HHS) through the Food and Drug Administration (FDA) and the Centers for Disease Control and Prevention (CDC) play an important role in promoting and ensuring food safety and also in promoting a nutritious diet for all Americans, especially in younger schoolchildren.

However, the recent outbreaks of food-borne disease due to bacterial contamination, while rare, point to the vulnerability of the food supply, whether domestically produced or imported. The U.S. food supply has become increasingly diverse in origin; as this trend continues, protecting the public from both known and potentially new sources and kinds of bacterial and food-borne illnesses will be an increasing challenge.

A continued public investment in food safety requires integrating the newest scientific and technological discoveries from the health sciences, developing new detection technologies, and a deep understanding of the entire process of food production, from the environmental conditions on the farm or ranch, through any possible exposure opportunity in food processing and distribution. Research opportunities also include continued investment in the regulatory science that supports the regulatory framework applied by the Food Safety and Inspection Service, as well as the FDA and the Department of Homeland Security (in cases of potential deliberate contamination). The goal of these efforts is to prevent food contamination from any source, safeguard livestock health, limit potential for zoonotic disease transmission, and build the evidence base for regulatory decision-making to protect and promote the Nation's public health.

In terms of nutrition, two related major challenges for the United States and for the world are the epidemics of obesity and diabetes. According to the CDC, more than one-third of U.S. adults are obese, another third are overweight, and 17 percent of U.S. children were obese in 2009–2010.[37] As a consequence, diabetes in the United States is increasing at an alarming rate; from 1980 through 2010, the percentage of people with diagnosed diabetes increased between 125 to 200 percent, depending on the age group.[38] This epidemic affects not only the health of the U.S. population, but also the entire U.S. economy due to increasing healthcare costs and a decline in the productivity of its workforce.[39]

How is this tied to agriculture? Essential components for combating the obesity epidemic are consumer education and access to nutritious food choices at affordable prices; it is also important to improve or augment the nutritional value of widely consumed foods for both domestic and international consumers. Much of the U.S. commodity crop yield is not used directly for human consumption, but rather used to provide derived food, either through animal production or processed food. Such food can be high in sugars, starch, and fat and low in protein and other nutrients. Nutritious food, such as fruits, vegetables, and lean meats, can be more expensive than processed foods, but there are also challenging social science issues that underpin consumer choice and preference.[40]

Currently, USDA nutrition guidelines focus strongly on lowering and controlling fat and salt intake, yet there is a wealth of accumulating scientific evidence that sugar should also be limited, especially in light of the diabetes epidemic: "The open policy question for agriculture and food policy is how to more widely and cheaply grow healthy crops and healthy sugars and limit highly processed, diabetes-inducing sugar."[41] As part of the public investment in agricultural research, funding should be provided to explore how plant and animal products can be used or modified to respond to this crisis in nutrition and health by developing new varieties of food products and new approaches to food processing. Historically, food processing and technology research was exclusively the domain of the private sector. But given the nationwide public health crisis presented by the obesity epidemic, there is a clear public need to understand how food, nutrition, and food processing can contribute to or alleviate this nationwide epidemic. Public-private partnerships are needed to translate and develop the connections between nutrition research, the box of food on the grocery store shelf, consumer choices, and health outcomes.

Feeding the World

The seventh major challenge facing agriculture is complex: The United States currently produces more than enough food for its own population. Indeed, over the past decade, roughly one third of U.S. corn production has been used for ethanol production rather than as food for livestock or people (corn production has increased in response to demand for ethanol). Partly due to the success of the agricultural enterprise, the area of land under cultivation in the United States has decreased from 4.2 million square kilometers in 1980[42] to 3.7 million square kilometers in 2010,[43] even with an increasing demand for bio-fuels. This suggests that the United States will be able to feed its own population over the next several decades, assuming that it is able to meet the various challenges discussed above. This has not always been the case; as recently as 1970, the Southern Corn Leaf Blight swept through 80 percent of the existing U.S. corn crop in 4 months, creating a huge food and national security issue.[44] We have been fortunate not to have such a crisis in more recent years, whether due to a new pest or pathogen or a heat wave like what Russia experienced in 2010. But the agricultural enterprise must remain vigilant to such disturbances because the United States depends on excess food production to support food security in the world and benefits economically from strong food exports. Western Europe is in a similar situation, as are many other countries in the developed world.

Unfortunately, much of the rest of the world does not share our good fortune. A July 2012 report projects that within the next decade the proportion of the world population that will be food insecure will decline from 24 to 21 percent, but the total number of food-insecure people (given world population growth) will increase by 37 million to nearly 900 million people worldwide, with the greatest severity in sub-Saharan Africa.[45] Because of growing population and changing diets (including increased demand for animal protein in emerging economies), global food production must roughly double (increase 100 percent) over the next 40 years to meet global demand if current trends in population growth and dietary choices continue.[46] Yet simulations by the USDA's Economic Research Service (ERS) project increases in global average farm productivity and increases in food of only 40 percent, defining the so-called "productivity gap" between need and output.[47] Even this may be an optimistic view, as it does not consider the challenges of a changing climate. Failure to produce enough food may create security issues related to social unrest generated by food shortages, especially in the more volatile parts of the world, such as those that triggered the 2011 "Arab Spring."[48,49] The United States has a strategic and security interest in maintaining a strong global food market and avoiding food shortages, especially in regions that are already politically or socially unstable. As President Obama remarked to the G8 leaders on May 18, 2012,[50] "We've seen how spikes in food prices can plunge millions into poverty, which, in turn, can spark riots that cost lives, and can lead to instability. And this danger will only grow if a surging global population isn't matched by surging food production. So reducing malnutrition and hunger around the world advances international peace and security—and that includes the national security of the United States."

The U.S. agricultural research enterprise has two roles in helping alleviate the growing problem of international food security. The first is to invest in innovation that will increase the efficiency and intensity of food production in the developed world with respect to average yields per land area. But there is a question of whether the United States and other developed countries can continue to increase yields at the level of USDA projections. There are also significant scientific concerns about how far crop yields and livestock and livestock-derived productivity (milk, eggs) can be pushed by biological improvements from marker-assisted breeding or technological improvements from precision agriculture in the face of inherent physiological barriers and environmental limitations.

Increases in food production should also come from the development of nontraditional or newly domesticated crops, especially in marginal lands or in more temperature or water-stressed conditions. Of the more than 50,000 edible species of plants, we currently exploit fewer than 50, with 15 plants supplying 90 percent of the world's food, and three crops rice, maize, and wheat providing 60 percent of the total.[51] To utilize species that are currently under-exploited, however, would require (1) research to understand their basic biology; (2) collecting, maintaining, and characterizing existing genetic diversity to envision how they might fit into existing or potential production systems; and (3) undertaking long-term, vigorous efforts to improve that crop or livestock strain for optimal performance.

The research needed to address these concerns about increasing food production requires a blend of private and public funding. Critical points of emphasis include developing new crop varieties and reducing the fertilizer and pesticides required to maintain high yields. Many of these aspects will not be addressed by private-sector research alone. Some of the necessary scientific research is too basic and too far from direct product development, such as exploring and domesticating new food crops. No single company will fund such research because the expected results are too general and thus too difficult to monetize by the initial investor.

A second role for the U.S. agricultural research enterprise in dealing with the challenge of global food security, and that may be most important in actually achieving these long-term goals, is to assist countries in developing their own agronomic practices that will increase yields while minimizing environmental impacts. Such investments will be particularly important in the poorest regions of the world, where yields of major crops like corn or wheat are as much as 10 times lower per unit land area than in the United States or in Western Europe.[52] These research and outreach efforts, which are squarely in the public domain, are supported by cooperative efforts between the USDA and the Department of State through the U.S. Agency for International Development (USAID). Investments in research aimed at assisting the world's poorest people and minority farmers and women, with their agriculture will not only lead to greater food stability and security, potentially reducing our military expenditures in the future, but are also likely to result in other valuable outcomes such as decreased deforestation and local economic development. Note that private foundations have been active in this area and have stimulated significant agriculture developments, such as the effort by the Gates Foundation, in partnership with the NSF's Basic Research to Enhance Agricultural Development (BREAD) program, to advance basic research on key problems in small farmer agriculture in the developing world.

Summary, Integration, and Big Data

PCAST believes the seven scientific "grand challenges" for agriculture just described represent the most pressing and urgent challenges for agricultural research. We firmly believe that public investors in agricultural research at the Federal, state, and local level should organize R&D agendas around these challenges.

Many of the challenges described earlier are highly interconnected. For example, addressing water use in agriculture should not be done independently of considering water quality and the efficient use of fertilizers and pesticides. Research on new pests and pathogens must simultaneously consider the impacts of a changing climate. As the agricultural research enterprise tackles these challenges, there will be a greater need for integration and synthesis.

Along these lines, one overarching challenge is the need for better information technology capabilities. Modern technology allows for the collection and use of many different types of agricultural data, from soil moisture and chemistry, meteorology and market conditions, crop and market conditions, consumer nutrition and preference, to gene sequences and ecological variables. Data sets in many of these fields are massive, which presents challenges for accessibility, interoperability, and persistence. As research efforts proceed, there will be a need for better data-management strategies addressing such issues as data storage, search algorithms, analytical methods, data sharing, and data visualization. Other research communities, including medicine and energy, are struggling with similar issues, and there are good opportunities for collaboration with DOE, NIH, NSF, and their global partners in this effort.

Finally, we note that the challenges listed earlier overlap with previously identified high-priority areas within publicly funded agricultural research. For example, USDA's Agriculture and Food Research Initiative, home of the competitive grants program, has five explicit goals: keep American agriculture competitive while ending world hunger; improve nutrition and end child obesity; improve food safety for all Americans; secure America's energy future through renewable fuels; and mitigate the damage by, and adapt agriculture to, variations in climate. The hurdle going forward is not just identifying the looming challenges facing agriculture, but to continually optimize and focus the entire agriculture research enterprise on solving these grand scientific challenges.

III. CURRENT STATUS OF THE AGRICULTURE RESEARCH ENTERPRISE

In the United States, agricultural research comprises a wide range of Federal, state, and private sector activities that range from understanding the basic biology of living systems to the high-tech development of precision agriculture. These activities are funded by a complex mixture of Federal and state agencies, the private sector through industry and venture capital, philanthropic foundations, and programs funded by agricultural producers who contribute portions of their profits through statutory "check-off" or voluntary programs. The Federal investments alone span a continuum from fundamental research in a wide array of agriculture-related disciplines to applied innovations focused on increasing production of food, feed, materials, and energy and on food safety and nutrition. These efforts yield important societal returns. A large body of economic literature, including 35 studies published over the time period of 1965–2005, indicates that the median estimate of the social rate of return was 45 percent per year and that for every $1 spent on agricultural research, approximately $10 worth of benefits were returned to the economy.[53] Assessing the scope of the U.S. agricultural R&D enterprise is not a trivial matter.[54] After considering the publicly available information, we chose to take a broad view of the scope and the players involved in funding and performing U.S. agricultural R&D (see box below). We included both public and private funders, given that private expenditures represent 61 percent of total spending for agricultural R&D in 2000 (Figure 1).[55] Finally, estimating the total dollars spent on agriculture research presents a challenge due to the multiple agencies and funders, the global nature of agriculture, and whether or not specific activities are deemed as agricultural research. This complex landscape of funding, organizations, and research both presents opportunities and represents liabilities. Opportunities include a broad base of support for agriculture, with multiple stakeholders representing different interests and expertise. Liabilities include a lack of coordination, lack of agreed-upon priorities, and potential overlap and gaps in both funding and research (between the public and private sector and within public funding itself).

Agricultural Research & Development

Agricultural R&D includes a diverse set of scientific and technological disciplines that support the production of food for human or animal consumption within existing and future resource limitations. Public agricultural R&D largely focuses on the plants and animals used in agricultural production; the natural resources that are used or affected by agriculture production; production systems and their support; food and nonfood products; human nutrition, wellbeing, and food safety; families and community systems; economics, marketing and policy; and research administration. Private R&D includes many of the same research topics, specifically plant breeding, agricultural chemicals, farm machinery, fertilizers, animal health, animal breeding, animal nutrition, and food and kindred products. This report broadly defines agriculture R&D to also include demonstrating existing or improved agricultural practices (extension services), as well as gathering agricultural statistics that are critical to the agricultural research mission, biofuels research, and forestry R&D funded by the U.S. Forest Service.

A. Research Funding for Agriculture in the United States[56]

Figures 1 and 2 describe funding of U.S. agricultural research. Figure 1 shows that U.S. public and private research in agriculture and food totaled more than $14 billion in 2009, of which $3.8 billion was Federal funds. R&D in the private sector accounted for the majority (61 percent, $8.7 billion) of research funding. Of this $8.7 billion, 89 percent went to industry-managed internal research, and 11 percent went to land grant universities, other research and comprehensive universities, and State Agricultural Experimentation Stations (SAESs).[57] State funding accounted for $1.9 billion disbursed to land grant universities and experiment stations; most of this $1.9 billion is required state matching funds that must be appropriated to receive concomitant Federal funding.

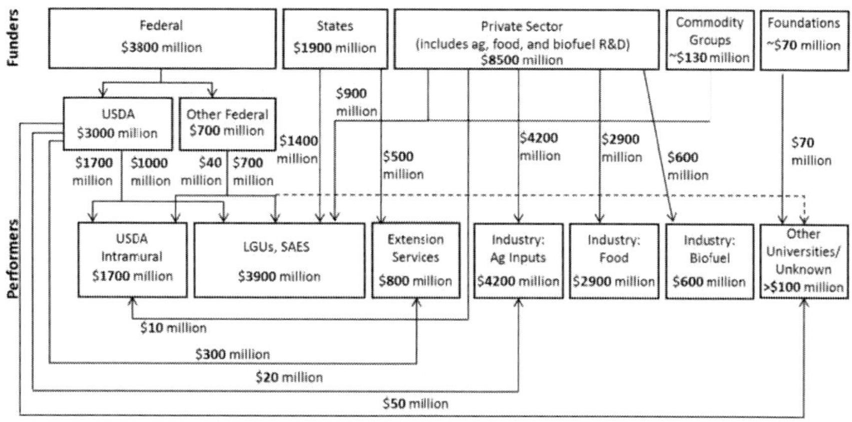

Notes: "States" represents state funding of land grant universities (LGUs), State Agricultural Experimentation Stations (SAES), and extension services. "USDA" represents total USDA funding of Research, Education, and Economics (REE) budget for FY2009, and "USDA Intramural" represents research performed by the Agricultural Research Service, Economic Research Service, Forest Service, and National Agricultural Statistics Service. "Other Federal" represents all non-USDA Federal agency funding. "LGUs, SAES" represents non-extension funding from states, Federal sources, and additional nonpublic entities, including commodity groups. "Private Sector" represents all private U.S. research funding for agricultural inputs sector ("Ag Inputs"), the food manufacturing sector ("Food"), and the biofuels sector ("Biofuels"). "Biofuels" values have been adjusted to avoid double counting funding already accounted for under "Ag Inputs" companies. "Commodity Groups" represents the 18 Federally authorized check-off programs. "Other Universities/Unknown" represents non-landgrant universities and any research funding where performers are unknown. "Foundations" represents agricultural R&D funding from major U.S. foundations. The $20 million estimate from USDA to industry represents Small Business Innovation Research (SBIR) funding. Figures rounded to the nearest $100 million except those below $100 million. Totals may not sum due to rounding.

Sources: Public funders and public performers from Current Research Information System (CRIS), USDA REE budget, and USDA Plan of Work for extension services. Extension services funding includes both Federal outlays from USDA and state matching funds as reported by LGUs to USDA but excludes $630 million in self-reported expenditures from sources of unknown origins that may overlap with other shown sources. SBIR estimate from National Science Foundation, National Center for Science and Engineering Statistics, "Federal Funds for Research and Development," Table 8. "Ag Inputs" and "Food" estimates from USDA, Economic Research Service (ERS), "Research Investments and Market Structure in the Food Processing, Agricultural Input, and Biofuel Industries Worldwide." Data for certain sectors come from years other than 2009, and the latest data available were utilized. "Industry (Biofuel)" estimate also from ERS report, but updated based on personal communication with USDA. "Commodity Groups" estimate from PCAST survey of commodity groups. "Foundations" estimate from USDA personal communication. Flow from "Private" to "LGU-SAES" from CRIS and added to ERS data after removing spending by commodity groups, which was assumed to be included in CRIS. See Appendix A for description and limitations of each source and combined sources.

Figure 1. 2009 U.S. public and private agricultural research, development, and extension expenditures.

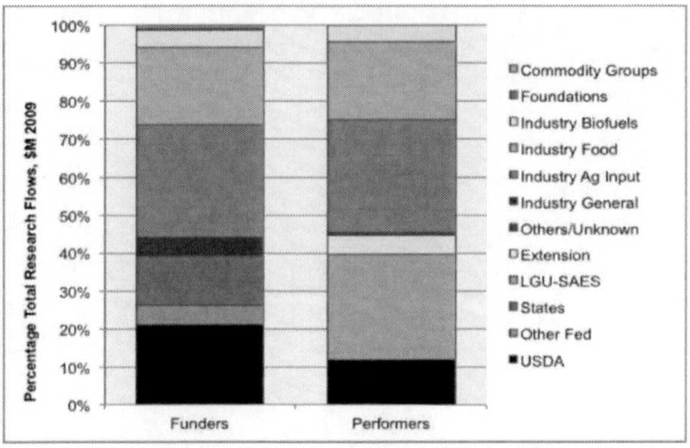

Public entities fund 39 percent and perform 46 percent of agricultural research. Private entities fund 61 percent and perform 54 percent of agricultural research.

Notes: "States" represents state funding of land grant universities (LGUs), State Agricultural Experimentation Stations (SAES), and extension services. "USDA" represents total USDA funding of Research, Education, and Economics (REE) budget for FY2009, and "USDA Intramural" represents research performed by the Agricultural Research Service, Economic Research Service, Forest Service, and National Agricultural Statistics Service. "Other Federal" represents all non-USDA Federal agency funding. "LGUs, SAES" represents non-extension funding from states, Federal sources, and additional nonpublic entities, including commodity groups. "Private Sector" represents all private U.S. research funding for agricultural inputs sector ("Ag Inputs"), the food manufacturing sector ("Food"), and the biofuels sector ("Biofuels"). "Biofuels" values have been adjusted to avoid double counting funding already accounted for under "Ag Inputs" companies. "Commodity Groups" represents the 18 Federally authorized check-off programs. "Other Universities/Unknown" represents non-landgrant universities and any research funding where performers are unknown. "Foundations" represents agricultural R&D funding from major U.S. foundations. The $20 million estimate from USDA to industry represents Small Business Innovation Research (SBIR) funding. Figures rounded to the nearest $100 million except those below $100 million. Totals may not sum due to rounding.

Sources: Public funders and public performers from Current Research Information System (CRIS), USDA REE budget, and USDA Plan of Work for extension services. Extension services funding includes both Federal outlays from USDA and state matching funds as reported by LGUs to USDA but excludes $630 million in self-reported expenditures from sources of unknown origins that may overlap with other shown sources. SBIR estimate from National Science Foundation, National Center for Science and Engineering Statistics, "Federal Funds for Research and Development," Table 8. "Ag Inputs" and "Food" estimates from USDA, Economic Research Service (ERS), "Research Investments and Market Structure in the Food Processing, Agricultural Input, and Biofuel Industries Worldwide." Data for certain sectors come from years other than 2009, and the latest data available were utilized. "Industry (Biofuel)" estimate also from ERS report, but updated based on personal communication with USDA. "Commodity Groups" estimate from PCAST survey of commodity groups. "Foundations" estimate from USDA personal communication. Flow from "Private" to "LGU-SAES" from CRIS and added to ERS data after removing spending by commodity groups, which was assumed to be included in CRIS. See Appendix A for description and limitations of each source and combined sources.

Figure 2. Total U.S. agriculture and food research, development, and extension expenditures by research funder and performer for 2009.

Federal Agencies

Federally funded agricultural R&D accounted for approximately 26 percent of the total agricultural and food research in the United States in 2009, the last year for which all data sets are available (Figures 1 and 2).

1. USDA

The USDA, appropriately, is the largest Federal funder of agriculture R&D (Figures 1 and 2), both for university and government researchers. Of the total Federal expenditures on agricultural R&D to universities in 2009, USDA provided more than half the total funding for agricultural research from its extramural programs (Figure 3).[58] Through a variety of funding mechanisms (including formula funds, noncompetitive grants, and competitive grants), $1.4 billion, is awarded extramurally to land grant universities, other research universities, SAESs, other cooperating institutions, and cooperative extension[59] representing approximately one-third of the USDA research budget (see Figure 4).[60]

2. Other Federal Agencies: DOE, NSF, HHS, EPA, USAID, DOD

Other Federal agencies also provide critical and substantial investments to agricultural research. The total agriculture-related R&D investment to land grant universities, other research universities, and SAESs by all other Federal agencies (e.g., National Science Foundation, Health and Human Services, Department of Energy, and Environmental Protection Agency) was $700 million in 2009 (see Figures 1 and 2). USDA also has a large intramural R&D program; 66 percent[61] of USDA's research funding was disbursed to intramural sources in 2009 including ARS, ERS, and FS (see Figure 4). Note that this is a substantial divergence (roughly double) from other Federal research-focused agencies, which devote 30 percent or less of their R&D budgets to intramural research (Figure 4). The Department of Energy (DOE) invests in energy crops and cropping systems, renewable/bioenergy technologies, relevant ecosystem and Earth-observation sciences, systems biology, and engineered innovation at the water/energy/land/agriculture nexus. The NSF supports an interagency plant genomics program devoted to crop plants and other basic research in a host of related disciplines across the life and physical sciences. The National Aeronautics and Space Administration (NASA) invests in relevant Earth observations and natural resources assessment.

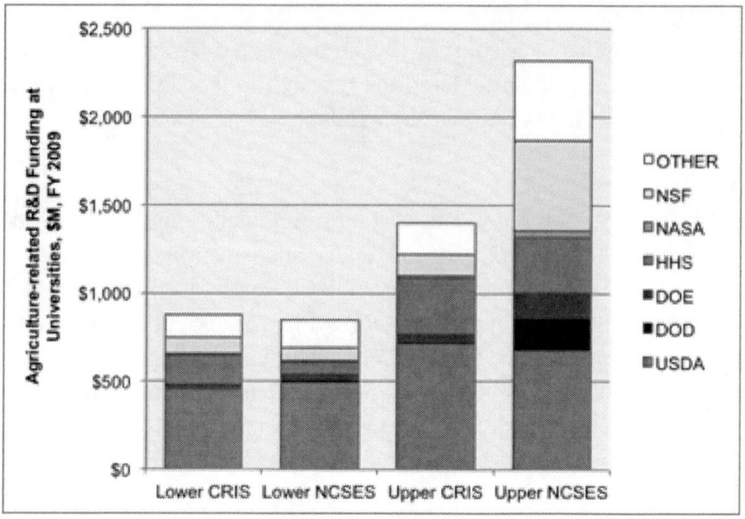

USDA funds approximately half of agricultural research at universities (Lower CRIS, Lower NCSES, Upper CRIS), but funds a smaller share of more broadly related research that includes basic plant and biological research, as well as biofuels (Upper CRIS). Two data sources (CRIS and NCSES) were used to assess agriculture-related research funding to universities, and final totals depend on the definition of agricultural research and the data source. The lower bound includes only funding for agricultural R&D, but the upper bound includes both agricultural and biological R&D. Note that there is some spillover and crossover between basic biological biomedical research and basic plant and animal research. Thus, the Upper CRIS data should only be interpreted as agricultural research related to, but not explicitly defined as, agricultural research.

Notes: "FFRDC" is a Federally Funded Research and Development Center. "Intramural" includes costs associated with the administration of intramural and extramural programs by Federal personnel and actual intramural performance. "Other" includes non-FFRDC nonprofits, foreign performers, and state and local governments.

Source: National Science Foundation/National Center for Science and Engineering Statistics, "Survey of Federal Funds for Research and Development: FY 2008–10." See Appendix A for description and limitations of source.

Note that the data presented in Figure 3 were collected separately and from different data sources than the data presented in Figures 1 and 2. Therefore, the USDA data represented here are not entirely consistent with Figure 1 and 2. However, the NCSES data here are collected and reported using a standardized method for all Federal agencies, which thus provides a uniform, comparable view of different agencies.

Figure 3. Federal funding of agriculture-related research to universities is between $850 million and $2.3 billion.

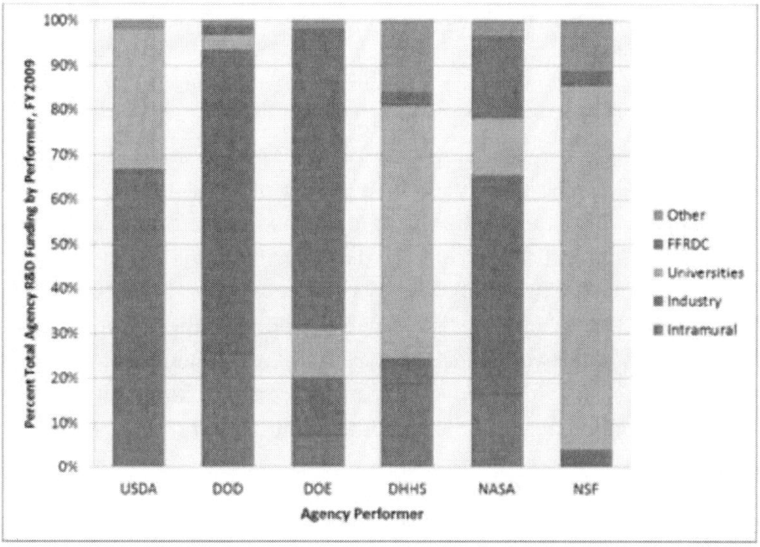

Intramural R&D at USDA accounts for 66 percent of the R&D budget. Total R&D funding of agencies: USDA at $2.3 billion, DOD at $68 billion, DOE at $9.9 billion, DHHS at $36 billion, NASA at $5.9 billion, and NSF at $6.1 billion.

Notes: DOD is Department of Defense. DOE is Department of Energy. HHS is Department of Health and Human Services. NASA is National Aeronautics and Space Administration. NSF is National Science Foundation. "Lower NCSES" includes agricultural science R&D funding only, which is assumed to not fully capture basic biological science. "Upper NCSES" includes agricultural science and biological science R&D funding. "Upper NCSES" estimate from HHS is adjusted to equal the "Upper CRIS" estimate to account for the large overestimate of agriculture-relevant R&D stemming from medical research. "Upper CRIS" represents full estimate of agency R&D expenditures from CRIS data, and "Lower CRIS" removes rural development, agricultural engineering, food systems, economics, nutrition, and basic plant and animal biology to compare with the definition from NCSES agricultural science (Knowledge Areas 400900, 206, 304, and 305 in CRIS). NCSES and CRIS data exclude extension services.

Sources: Lower and Upper CRIS estimates from Current Research Information System (CRIS). Lower and Upper NCSES estimates are from National Science Foundation's National Center for Science and Engineering Statistics (NCSES), "Survey of Research and Development Expenditures at Universities and Colleges," FY 2009. See Appendix A for methodology and a description and limitations of each source.[62]

Figure 4. Distribution of the R&D portfolio across Federal funders of agricultural research (as listed in Figure 3).

The Environmental Protection Agency (EPA), the USAID, and the Department of Health and Human Services (DHHS), through the FDA and the NIH, invest in diet-related research, public health, nutrition, food security, food safety, pesticides and related chemicals, as well as associated regulatory science for their respective regulatory activities. The Department of Homeland Security holds responsibility for security and protection of U.S. agricultural operations, food, and water supplies from intentional criminal acts. USAID with USDA is currently overseeing the distribution of funds from the President's Feed the Future Initiative, which focuses on global food security via innovation that promotes "sustainable intensification" of agricultural production with improved outcomes in human dimensions, particularly poverty alleviation and food and nutritional security. The DOD funds some programs in agriculture as it relates to global security and development.

State Support for Agricultural Research

State support for agriculture is relatively small and declining but still a critically important component of total public support. State government expenditures totaled $1.9 billion, with most of this funding dedicated to support State Agricultural Experiment Stations (often associated with Land Grant Universities) as a required state match to appropriated Federal formula funds. Indeed, two-thirds of the total University of California Experiment Station's budget was historically funded through state support. In the last few years, however, state support has declined throughout the United States[63] and now makes up less that 15 percent of the total agriculture research enterprise (Figure 2). States also provide support for the Extension Service, which serves as an important interface between the land grant universities and farmers and provides farmers, industry, and citizens with information and advice based on university research. Pressures on state budgets have led some states to substantially reduce funding for Extension, which creates a gap in delivery of information to farmers. As a result, in many active areas, other agricultural input suppliers are substituting for public institutions as sources of extension advice.

As Federal support remains flat and state support declines, experts note that a shift in the use of Federal resources can occur with Federal dollars more frequently directed toward gap-filling, short-term, local agricultural problems with consequent erosion of longer term and enabling investments. While emergency Federal measures are sometimes necessary, it remains vital for states to support local and regional agricultural research activities, with Federal support focused on regional or national issues.

Private Sector

The private sector constitutes the majority of annual agricultural R&D spending in the United States. The two major private R&D sectors are agricultural inputs (e.g., commodity crop seed and biotechnology, crop chemicals, livestock, farm machinery, etc.), estimated at $4.2 billion and food manufacturing and processing, estimated at $2.9 billion. Along with $600 million for biofuels research, private sector funding amounted to $8.5 billion in 2009 (Figure 1). The private sector also funded $800 million of R&D performed at the land grant universities. Although private sector funding of research at land grant universities has played a critical role in recent years as state funding has declined, some critics have noted that this can sometimes lead to tensions between academic researchers and their private funders when research outcomes differ from industry goals. This emphasizes the crucial role of continued public funding in the overall agricultural research enterprise.

1. Large corporations

Private sector investment in agricultural input R&D continues to be dedicated primarily to commodity and high-value crop production and specifically to the development of improved seeds and crop protection chemicals for the most lucrative global markets, increased yields, and improved resource-use efficiency. In 2011, the six largest multinational companies with significant agriculture focus invested nearly $6 billion globally in R&D for these two product categories (commodity and high-value crop production), representing an average level of funding just above 10 percent of sales in these product categories.[64,65,66] Over the past several decades, the private agriculture sector has experienced substantial market consolidation across the food-supply system. For example, the eight-firm concentration ratio, that is, the current market share covered by the eight largest firms in the farm machinery, crop seed and traits, animal health, and crop-protection chemicals sectors, are 61, 63, 72, and 75 percent, respectively, for 2009.[67]

2. Small corporations

Private investment in crop production research beyond large corporations is small by comparison. A handful of companies, such as Mendel Biotechnology, Inc. and Ceres, Inc., conduct basic plant biology research and work to develop improved dedicated bioenergy crops. The total annual R&D investment of these companies and other smaller startups approaches $50 million worldwide for biofuel R&D.[68] In most cases, such small companies exist for a relatively short period of time, usually less than 10 years, during which projects reach the proof-of-concept stage, at which point they either fail or are acquired by a large, established company. There are estimated to be fewer than 60 small agricultural biotech companies in the United States with more than a handful of employees, and only 5 of these have been in existence for more than 10 years. In contrast, there are approximately 800 small to medium-sized biomedical biotechnology companies in the San Francisco region alone.

3. Venture capital and Other Strategic Investments

Small agricultural biotech companies are usually financed with a mixture of venture capital and "strategic investment" by large companies from a diverse set of sectors. Between 2001 and 2011, approximately $7.5 billion was reportedly invested in equity stakes in small agricultural biotech companies in the United States. The average equity investment was approximately $11.4 million, enough to support approximately 45 full-time-equivalent years (i.e., 10 people for 4.5 years).[69] Note, however, that many of the investments are in biofuels, which are more related to energy than agriculture investing and have seen substantial increases in the last six years. The biofuels/bioenergy mandates and subsidies created a demand for innovation that previously was nonexistent and that may be a useful future lens to stimulate core agriculture investing. Although independent venture capital firms have been increasing their investments in the life sciences and energy areas over the past two decades, their support of core agriculture-focused investments has been anemic. Between 2005 and 2011, U.S.-based venture capital firms invested $404 million[70] in the agriculture and forestry sector out of about $209 billion, which is just 0.2 percent of this sector. As a global marker of the current landscape of the agricultural innovation ecosystem, there were no initial public offerings (IPOs) from agriculture-related companies between 2005 and 2011, suggesting that a crucial generator of capital, IPOs—and a major route by which venture capital recoups its investment—is difficult to create in this sector. Until there are one or two headline venture capital successes in the agricultural sector, it will be hard to generate substantial venture capital activity. There are significant hurdles that limit the ability of venture capital to take on the fully developed role it does play in other sectors such as information technology or health care. In many sectors of agriculture, strategic funding from large companies has been a much larger component of financial support for small companies than independent venture capital equity investments. Sometimes this comes in the form of strategic equity investing, and in very recent years, there has been greater involvement from the venture-capital arms of large firms, such as Dow and Pioneer. One way to bridge these challenges for agriculture is to mobilize research universities to play a larger role in agriculture-related technology transfer. Some expert observers note opportunities for land grant universities and other research universities to form regional hubs and centers of excellence that can stimulate more activity from the lab to the marketplace. The energy field of the past seven years is a potential model for agriculture of the symbiotic relationship between universities, government involvement, and venture capital investing. By 2005, the energy challenge became clear in general society, and the energy opportunity therefore became clear in many leading universities. This spurred greater interest by entrepreneurial students and faculty and led many universities to redouble their efforts. Today, a high proportion of venture capital investments in energy are in new private commercial opportunities spun directly out of university laboratories.[71] Also, from 2005 to 2011, energy investing by venture firms increased from $320 million to $2.9 billion a year, or from 1.3 percent to 9 percent of U.S. Venture Capital investment dollars per year,[72] propelling energy to the third largest investment sector, just behind information technology and biotechnology.[73]

Farmer and Commodity Groups

Farmer-funded and farmer-controlled commodity “check-off” programs fund a substantial amount of agricultural research, mostly at land grant universities, through national, state, and local commodity boards that legally obligate producers to contribute a percentage of their profits back into the shared checkoff programs for future product-specific R&D. There are 18 Federally authorized checkoff programs and several that continue as coalitions of state authorized programs. Both short-term and long-term research are funded by national, state, and local commodity boards. In 2012, total funding from these combined programs for research purposes at the national checkoff level exceeded $130 million, although this is probably an underestimate and does not accurately represent many state and local checkoff activities. The commitment of the commodity boards to research is often influenced by the enacting legislation, the nature of the commodity, and the value of the commodity. Funding ranges from less than $500,000 for blueberries to almost $61 million for soybeans.[74] The national commodity boards tend to largely fund research on new use and product development research, while state commodity boards tend to fund more production-oriented research with an emphasis on problems that are unique to a given state or region. The majority of commodity board-funded research includes an extension and outreach component to enable the farmers funding the research to benefit as quickly as possible. Hence, these are not typically venues for long-range research.

Table 1. Checkoff program research funding by product in thousands of dollars

Research Funding Category ($Thousands)					
Product	**Nutrition & Health**	**Production**	**New Uses**	**Food Safety & Disease**	**Total**
Almond	1,500	1,500		1,500	4,500
Beef	940		845	1,400	3,185
Blueberry	421				421
Cherry	285	135			420
Citrus		3,100		5,500	8,600
Corn		500	500		1,000
Cotton					11,000
Dairy					~10,000
Eggs	2,000				2,000
Honey	80	200			280
Lamb					100
Lumber		1,200			1,200
Mushroom	425		300		725
Peanut		1,374			1,374
Popcorn					0
Pork		9,500	500		10,000
Potatoes			750		750
Sorghum		1,400			1,400
Soybeans		34,447	26,533		60,980
Sugar Beets		1,300			1,300
Wheat		6,250	6,250		12,500
Total	**5,651**	**60,906**	**35,678**	**8,400**	**131,735**

More than $130 million in funding is available for crop-specific research through checkoff programs. Checkoff programs are mandatory farmer contributions to the state or national checkoff program as a percentage of crop sold; these funds are used for a variety of purposes, including education, marketing, and consumer research. The data presented here only account for scientific research funded by the checkoff program in four broad categories: Nutrition & Health, Production, New Uses, and Food Safety & Disease.[76]

Notes: Estimates represent national and some state totals. For some commodities, state and local numbers are not available. Hass avocado, mango, and watermelon had no data available and could not be estimated.

Source: Accessed July 1 at: www.rti.org/pubs/beach_porkcheckoff_final.pdf. Figure 28 and 210 display production research and new product data. Pork, dairy, and lamb did not respond to survey requests, so they are estimated. Pork checkoff data derived from 2005 economic analysis of the pork checkoff programs. Dairy numbers estimated at a small fraction of the $280 million entire dairy checkoff program. See Appendix A for description and limitations of source.

Nevertheless, although the amount of funding by commodity boards is relatively small in comparison to public institutions, agricultural commodity groups have a substantial impact on agricultural research well beyond their funding amounts. Most funding is leveraged by requiring matching funds from industry or the research institutions that receive the funds. In the case of soybeans, for example, the leveraged funds over the past 8 years have averaged about $3 for every $1 invested by the United Soybean Board.[75] Funding from commodity boards can influence research priorities for an institution, not only through grants to individual researchers but by funding facilities at land grant universities that are devoted to specific needs of the commodity groups. Although extremely useful, this mechanism is focused on short-term gains for a narrow, commodity-limited user group.

B. Agriculture Research Performers in the United States

As with funders, the performers of agriculture research are complex, occurring across many different types of Federal, state, and private institutions and companies. Below we discuss the institutions that conduct research and comment on the focus of their research.

Educational Institutions

University research and training generates agricultural innovation and the workforce to deploy those innovations. Land grant universities in particular provide a large portion (see Figure 2, right column) of today's agricultural research activities for the United States. Indeed, according to NCSES data, approximately 94 percent of the Federal funding for agricultural science R&D was conducted at land grant universities in FY09.[77] USDA distributes its funding for agricultural research at land grant universities through "formula funds" that are distributed to each SAES using a formula based on census data. These funds are subsequently distributed within the SAES-associated land grant university at the discretion of the SAES director. Some universities choose to distribute these funds competitively within their university, while others choose to use the funds for permanent faculty salaries. The latter option, where funds are guaranteed regardless of outcomes, could unintentionally dissuade researchers from the challenging work of performing at the cutting edge of science and ultimately producing novel, innovative research. Historically, some uses of these formula funds have been critiqued for not efficiently stimulating high-quality research.[78]

Other research universities provide a core expertise in basic science that underlies many of the emerging solutions to U.S. agricultural challenges. Agricultural research funding at other universities is almost exclusively through peer-reviewed competitive grants from a variety of Federal agencies as discussed earlier and, at approximately $100 million, is a tiny fraction of the research performed at land grant universities. One of the drawbacks of the current system of agricultural research is that there is often a separation of agricultural research from other areas of biology, chemistry, social sciences, earth sciences, computer sciences, and engineering. Although it is understandable that the land grant universities play a leading role in the agricultural research enterprise, it is essential that other research universities participate in the effort to address these challenges (described in Chapter 2), allowing the agricultural research enterprise to benefit from the substantial public investments in basic research in other areas. Indeed, 18 of the top 20, and all of the top 15, university recipients of NIH funding are not land grant universities,[79] suggesting little overlap between life sciences and agricultural research by institution. At times, this apparent fragmentation between agricultural research and sustained interaction with other basic sciences at the university level can perhaps prevent or delay the transfer of knowledge and discovery, ultimately delaying the agricultural gains that are needed.

Furthermore, universities of all types play an essential role in the agricultural research enterprise by training the undergraduates, graduate students, and postdoctoral researchers who will form the foundation of tomorrow's highly trained workforce. Land grant universities have additional mandates to disseminate their research findings to clientele—farmers, industry, and land managers—and the general public. This extension activity is the model for agriculture extension activities worldwide.

Programs focused at the high-school level have recently become a strategic focus, in recognition of the historic positive impacts of national organizations such as FFA (formerly known as Future Farmers of America) and activities supported by USDA Cooperative Extension's 4H programs. The FFA and the very successful "agricultural high schools" demonstrate that if students are exposed to agriculture in grades 9–12, increased enrollment in agriculture majors in college result. And there is great demand for such training. The oldest such school is the public Chicago High School for Agricultural Sciences (CHSAS), on the south side of Chicago and now in its 27th year.[80] The school draws from an urban population, and only a small percentage of students who apply via lottery for admission can attend due to a large application pool and space limitation.[81] Many CHSAS graduates are minority, first-generation college students and are now contributing to industry, government, and education, and some are now farming in urban or rural settings. At the same time, the growth in agronomy students from urban settings with little or no farm background creates other challenges. As budgets shrink, it is difficult to provide these students the meaningful, hands-on field and farm-scale experience that would be assumed for students who grew up on a farm.

Research at the USDA

Research performed at the USDA includes four intramural programs, the Agricultural Research Service (ARS), the Economic Research Service (ERS), the Forest Service (FS), and the National Agricultural Statistics Service (NASS), and one extramural program, the National Institute of Food and Agriculture (NIFA). NIFA, previously known as the Cooperative State Research, Education and Extension Service (CSREES), was formed in 2009 to foster research innovations in agriculture.[82] Although about half of NIFA funds are awarded to universities and state agricultural extension agencies through legislative formulas (based on population), and a substantial portion of other funds are small, congressionally mandated, noncompetitive projects, some portion of NIFA funds are competitively awarded to extramural research at both land grant and other universities. In this capacity, NIFA is the primary source of competitive grants at the USDA. Competitive grants within NIFA in Fiscal Year 2012 represent 16 percent of the entire research budget of USDA, an increase from 10 percent in 2005 (Figure 5). The total budget for NIFA is much larger than this competitive portion (approximately $1.3 billion, or 45 percent of total USDA research funding), but much of this is prescribed to formula funds for research, education, and extension and other noncompetitive programs. The Agricultural Research Service (ARS) is responsible for the largest proportion of intramural research at the USDA. ARS has a large scientific staff of roughly 2,200 people and conducts research at multiple locations across the country.[83] It maintains four regional centers as well as individual research stations, many of which are located on or near land grant university campuses. The Beltsville Agricultural Research Center in Beltsville, MD, is the largest and houses the National Agricultural Library, the National Plant Germplasm System, and the Human Nutrition Research Center. ARS represents the largest fraction of USDA research funding (37 percent in FY 2012; see Figure 5). Research at the ARS focuses on four national program areas: nutrition, food safety, and quality; animal production and protection; crop production and protection; and natural resources and sustainable agriculture systems.[84] Historically, a mission of ARS was to undertake long-term research that would not be done by the private sector or by research universities. As a service to the broader agricultural research community, ARS has the additional responsibility to maintain unique research infrastructure for agriculture such as the Long Term Agricultural Research sites and important resources such as crop, animal, and microbial genetic collections.

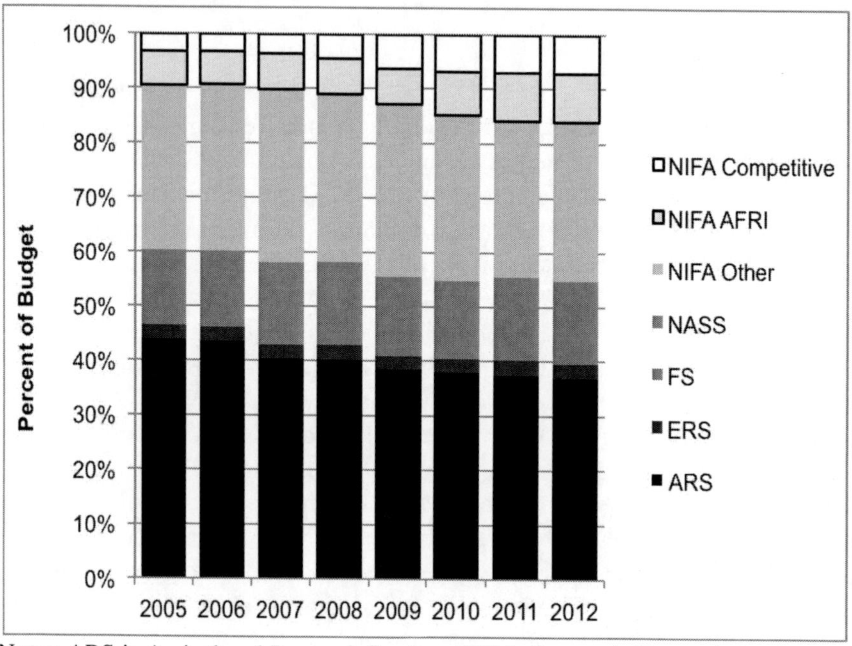

Notes: ARS is Agricultural Research Service. ERS is Economic Research Service. FS is Forest Service. NASS is National Agricultural Statistical Service. NIFA is National Institute of Food and Agriculture. NIFA Competitive and NIFA AFRI highlighted to represent percentage of competitive USDA funding. Before 2009, NIFA numbers represent Cooperative State Research, Education, and Extension Service (CSREES). The NIFA budget is divided into three categories: (1) "NIFA Competitive" is all non-AFRI R&D expenditures open to universities and nongovernmental entities in a competitive manner; (2) "NIFA AFRI" is the Agricultural and Food Research Initiative; and (3) "NIFA Other" is all other NIFA expenditures including formula funding.

Source: USDA Research, Education, and Economics (REE) and Forest Service budget information. See Appendix A for description and limitations of source.

Figure 5. USDA Research, Education, and Economics (REE) funding from Fiscal Year 2005 to Fiscal Year 2012.

The Economic Research Service (ERS) is the primary source of economic information and research in the U.S. Department of Agriculture. The ERS mission is to inform and enhance public and private decision-making on economic and policy issues related to agriculture, food, the environment, and rural development. ERS staff disseminates economic information and research results through an array of outlets, including agency-published research reports, market analysis and outlook reports, economic briefs, and data products.[85] The Forest Service R&D efforts range across the biological, physical, and social sciences to promote sustainable forest and rangeland management throughout the United States. Its work focuses on informing policy and land-management decisions and involves a range of partners, including other Federal and USDA agencies, academia, nonprofit groups, and industry. Forest Service R&D is conducted at over 67 field sites and 80 experimental forests and ranges.[86] The mission of the National Agricultural Statistics Service (NASS) is to provide timely, accurate, and useful statistics in service to U.S. agriculture. NASS conducts hundreds of surveys every year covering the production and supply of food and fiber, prices paid and received by farmers, farm labor and wages, farm finances, chemical use, and changes in the demographics of U.S. producers, among other areas.[87] The U.S. Cooperative Extension Service is a nationwide educational network, based largely at land grant universities and connected to a system of local and regional offices. Although the number of local extension offices has declined over the years, and some county offices have consolidated into regional extension centers, there remain approximately 2,900 extension offices nationwide.[88] Cooperative extension generally serves as the bridge between universities, the ARS, and local agricultural practitioners. University-based cooperative extension specialists and county-based cooperative extension educators (agents) conduct applied research and provide useful, practical, and science-based information to agricultural producers, small business owners, students, consumers, and others in rural areas.

Public Sector Research Areas

The combined public research portfolio is distributed across a diversity of crops and animals (Figure 6AB). Commodity crop research (corn, soy, wheat, cotton, and rice) constitutes 27 percent of public research funding and 36 percent of USDA intramural funding. The "other plant" category shown in Figure 6A represents over 119 additional plant species and crop varieties, including all fruits, vegetables, legumes (other than soy), fiber crops, grasses, garden crops, and tree nuts. Animal funding is rather evenly distributed across the major food animals preferred by the American public (Figure 6B).

Private Sector Research Areas

Crop production research in the private sector primarily focuses on production of staple crops such as corn, wheat, rice, soybeans, and cotton, as well as some large-scale vegetable crops. These represent the most widely grown and consumed crops globally, with well-defined markets and significant economic importance to the United States. Relatively little private investment is dedicated to the improvement of other crops with smaller market shares, including most of the fruits and vegetables consumed in the United States. One of the more pressing challenges and opportunities for private agricultural research is the simultaneous growth in the scientific disciplines involved in modern agriculture and in the quantity and complexity of biological and environmental data being generated. The scope and scale of scientific disciplines relevant to today's crop and livestock production have dramatically changed in the past two decades. Private agricultural R&D programs must balance their efforts between basic discovery research and product-development activities and technologies such as precision agriculture. With both crop seed and chemical productdevelopment costs and time lines expanding, the proportion of investment available for basic research is diminishing, particularly for those projects requiring long-term studies to receive regulatory approval.

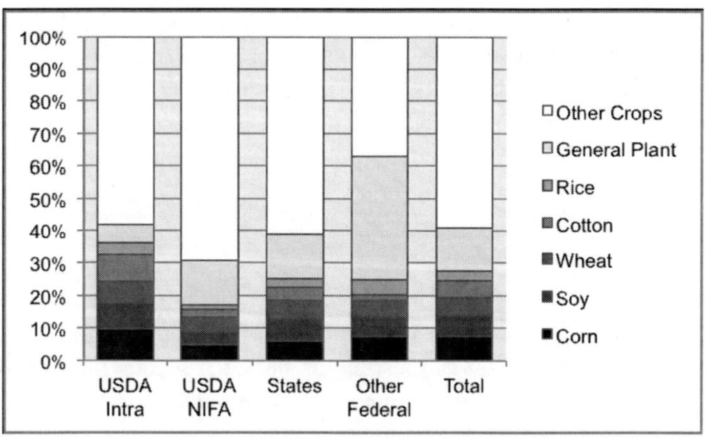

(A)

Figure 6. (Continued)

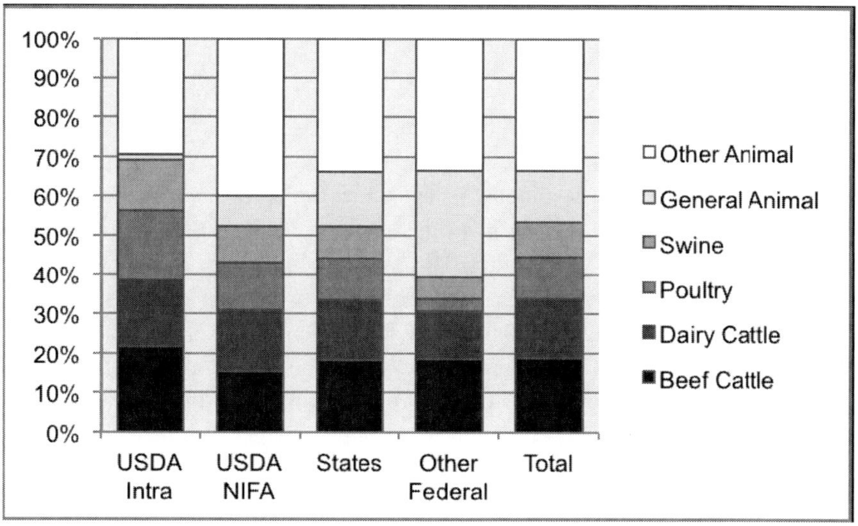

(B)

Notes: "General Plant" includes multiple crop research, general plant research, or non-crop plant research such as trees and ornamentals. "Other Crops" includes all other crops besides corn, soy, and wheat, including all fruits, vegetables, tree nuts, tobacco, sugar, other grains, other oilseeds, other fiber crops, grasses, garden plants, and specialty crops (over 119 categories). "General Animal" refers to research relevant to multiple animals or general animal research. "Other Animal" includes all other animals besides pork, poultry, dairy, and beef, including companion animals, fish, horses, goats, laboratory animals, bees, invertebrates, and sheep. "USDA Intra" is intramural expenditures by the Agricultural Research Service and Forest Service. NIFA is National Institute of Food and Agriculture. ARS is the Agricultural Research Service. FS is the Forest Service. NIFA is the National Institute of Food and Agriculture.

Source: USDA Current Research Information System data by subject of investigation. See Appendix A for description and limitations of source.

Figure 6. 2010 public agricultural R&D expenditures by crop (A) and animal (B).
Crop R&D expenditures totaled $1.4 billion and animal R&D expenditures totaled $1 billion.

Special Research Institutes

Although relatively few national research centers that involve collaborations between universities and government scientists have a focus on agricultural topics, a few have made important contributions. One example is the USDA Plant Gene Expression Center in Albany, California, which has a distinguished history of research in basic molecular biology of plants. (Approximately one-quarter of the staff scientists have been elected to the U.S.

National Academy of Sciences.) The U.S. Dairy Forage Research Center is a global leader in integrated research that improves livestock agriculture while innovating to reduce negative environmental consequences of livestock agriculture with particular focus on greenhouse-gas emissions. Another is the DOE Plant Research Laboratory at Michigan State University, which has been a leading center in plant biochemistry since the 1960s. These and other examples indicate that research institutes can be a highly productive way to create innovation in the agricultural system, especially if provided with core research funds to tackle long-term projects. An expansion of such research centers to focus on future challenges in the agricultural sciences across the land grant and other research universities would be one of the most efficient ways to encourage innovation and to provide consistent long-term funding commensurate with the long time lines of agricultural research.

C. Summary

Several compelling points emerge from the current status of agriculture research funding in the United States. First, funding for competitive research is low within the USDA (16 percent of research budget), although it has been growing in recent years (Figure 5). We discuss the implications of this finding at length in Chapter 4. Second, funding levels for USDA research has changed little since 1990, and public funding overall for agricultural research has changed little since the 1980s.[89,90] Third, the lack of a vigorous, competitive program in USDA handicaps the research effort upon which future developments in agriculture rest. Fourth, intramural research at USDA represents approximately two-thirds of the total research budget (Figure 4). Although ARS research is often dedicated to long-term, public goods, and is important for future preparedness, this large commitment may inhibit the growth of a vibrant competitive grants program. Finally, in response to stakeholder demands, USDA supports research on commodity crops that may have substantial overlap with the efforts of private industries and which dwarf the USDA funding commitments. Corn, soy, rice, wheat, and cotton account for 36 percent of the USDA intramural research budget and 27 percent of the public funding budget (Figure 6), raising questions about the appropriate allocation of research funds and whether they could be better spent on research challenges that are not a strong focus of the private sector. Public sector research also builds skills and trains the next generation of agricultural scientists. This training is a valuable public good but these skills are often transferable and can be developed through training on a wide variety of crops other than commodities.

IV. RECOMMENDATIONS: TOWARD AN INNOVATION ECOSYSTEM FOR AGRICULTURAL RESEARCH

This year we celebrate the 150th anniversary of President Lincoln's visionary legislation that founded the Department of Agriculture and the U.S. land grant university system. These institutions defined fundamentally new and different approaches to support the stabilization of democracy, the growth of the Nation, and its recovery from the Civil War. By focusing on agriculture and the development of a knowledge base, an educational system for the common person, and research capacity to improve agricultural outputs and outcomes, Lincoln supported the first transition of a rural population to the ranks of enfranchised citizen democrats, laying the foundation for the democratization of education that remains one of the country's greatest achievements.

Although vastly fewer U.S. citizens now engage in agriculture as a livelihood, the consequences of the choices we make in agriculture and in our food systems are arguably more significant now than in 1862. The U.S. food systems now have global impact in economic, environmental, security, and other dimensions, linking all parts of the world its own successes and failures and linking the United States to the consequences of agricultural failures in any other part of the world.

As discussed earlier, the U.S. agricultural research system is at a turning point. Decisions in the near future will determine how, when, and even if the United States meets the challenges to agriculture. Fortunately, there is good reason for optimism. Science is advancing at an unprecedented pace, with advances across the spectrum, from basic biological sciences, information sciences and engineering, to healthrelated sciences, presenting enormous opportunities for agriculture. With the application of new advances in science and technology, the opportunity for transformative change to U.S. agricultural systems with radical improvements in human, physical, and economic well-being, is near. But to accomplish the integration of scientific and technological changes into the agricultural system in a way that prepares the United States for the challenges ahead, we need to reconsider the current structure of agricultural research in the United States, reimagining our institutions and research programs for the next 150 years.

First, PCAST finds that the proportion of Federal funding for agricultural research allocated through competitive mechanisms is far below the proportion in other agencies, which fails to adequately encourage innovation.

Second, PCAST finds that the current agricultural research portfolio is not optimally balanced; areas that are identified as national challenges are in some cases underfunded while other pro-grams overlap with private sector activities.[91]

PCAST recommends the creation of a new innovation ecosystem for agriculture that leverages the best from different parts of the broad U.S. science and technology enterprise, focusing public investment on addressing the entire set of challenges to the agriculture enterprise. PCAST calls for a strategic investment that will create the path toward an improved innovation ecosystem for America's heartland, enhancing our economy, and harnessing the power of American innovation in science and technology.

Creating an innovation ecosystem for agricultural research requires investment in three main areas: research support, training and workforce development, and research infrastructure. In all three areas, PCAST has considered how to enhance and invigorate the existing structure while preserving the strengths and important roles of existing programs. In some cases, these reforms will require new funding to achieve the stated goals. In other cases, a restructuring of existing funds can be accomplished that enhances the use of competition for distribution of funding and expands the federal research portfolio to address additional research areas.

Overall, PCAST recommends that the United States increase its investment in agricultural research by a total of $700 million per year, focusing on addressing the emerging challenges described above.

In challenging budget times, PCAST recognizes the difficulty of recommending an increase in research funding. At the same time, PCAST feels that such an investment is appropriate, given the scale of the challenges and opportunities and the essential role that agricultural research has in the Nation's economy. In a recent report, the White House Council of Economic

Advisors observe, "The potential returns to increased research investment are particularly high today… These challenges will require supportive investment policies for basic and applied agricultural research to ensure that U.S. agricultural productivity continues to increase."[92] Moreover, the increased funding we recommend is a small fraction of the roughly $100 billion per year that the Federal Government spends on agriculture overall.

New, Refocused Investments in Agricultural Research

The United States will be unable to meet future challenges in agriculture without increasing its current investment in research. The USDA, in partnership with states and land grant universities, has a critical research mission to respond to local, short-term, and immediate issues that threaten current agricultural productivity, such as natural resource depletion or outbreaks of pests and pathogens. Addressing these local, immediate challenges is an essential service that the USDA and land grant institutions provide to the Nation, and responding to these challenges is essential to maintaining agricultural productivity. At the same time, it is important for Federally funded agricultural research to address the longer term challenges that have been outlined herein; indeed, long-term solutions to short-term problems will require sustained investment in basic research in agricultural science. We recommend an increased Federal investment in the agricultural research enterprise across multiple Federal agencies to address both the short-term and long-term research needs. Because the current agricultural research investment devotes insufficient funding and is not well-focused on the seven emerging challenges discussed in Chapter 2, this new Federal research investment should focus on them. We believe this will provide a powerful return on investment. As a 2010 National Research Council report notes, "One dollar invested in agriculture is one dollar invested in health, food, energy, and environment, as investments in agriculture are leveraged across these multiple areas."[93]

A new Federal investment in agricultural research should be accompanied by a rebalancing of the research portfolio to ensure that there are not substantial redundancies between research supported by private and public funds. This does not mean that research on major crops (i.e., corn, wheat, soy, cotton, and rice) should be eliminated from the Federal research portfolio. To the contrary, there are many areas of research on major commodity crops that are squarely in the public domain, including basic research, research on local and regional challenges to commodity crops that are neglected by industry, and research on some of the long-term challenges described above. But duplication of industry efforts is an inefficient use of limited resources.

PCAST believes that to stimulate innovation, distributing research funds through a competitive process should be at the core of our agricultural research enterprise. Over the last three decades, multiple independent review panels of experts have called for an increase in competitive funding of agricultural research. Indeed, a series of reports by the National Research Council (NRC), beginning four decades ago, assert that competitive funding will improve the quality of public funding for agricultural research; similar sentiments were expressed in 2004 by an independent task force report from USDA, mandated by Congress, which led to the creation of NIFA.[94] Some highlights from the NRC reports:

- From 1972: "that the Department of Agriculture seek a greatly increased level of appropriations for a *competitive* grants program, which should include support of basic research in the sciences…that underpin the USDA mission… The committee recommends further that this program be administered in such a way that research proposals are subjected to evaluation by peer panels of selected scientists."[95]
- From 1989: "This proposal presents a program to strengthen the focus of U.S. science on agriculture. The premise is that a judicious but substantial increase in research funding through competitive grants is the best way to sustain and strengthen the U.S. agricultural, food, and environmental system."[96]
- From 2000:"Without a dramatically enhanced commitment to merit--based peer-reviewed food, fiber, and natural-resources research, the nation places itself at risk." And, "Our recommendations reaffirm and extend the earlier Research Council vision for fundamental merit-based peer-reviewed research in food, fiber, and natural resources."[97]
- From 2003: "A realignment of the existing research budget to increase the proportion of funds in competitive grants and cooperative agreements would be effective in achieving greater flexibility and for addressing new and emerging issues by engaging new talent and expertise."[98]

Many of the advances in science and technology that led to new developments that enhanced the U.S. economy are the result of a significant

Federal investment in competitive, peer-reviewed research grants. Such programs are at the heart of the highly successful research enterprises supported by NSF, NIH, and DOE. Because of political pressures to support the large intramural program and the land grant universities, the competitive grants program within the USDA has remained far below what is required to support a vibrant, successful research enterprise prepared to meet the scientific challenges of the next few decades. In addition, the investment in basic science supporting agricultural research by other Federal agencies has been insufficient to meet the impending challenges.

The need for the adoption of a competitive process for awarding agricultural research funding applies not only to the USDA extramural research portfolio within NIFA, but also to funds for the intramural research program and for land grant universities. A larger fraction of intramural funding with the USDA could be awarded through a competitive process, much like the DOE National Laboratories or the NIH intramural program. Formula funds distributed to land grant universities could also be distributed within each institution through a more competitive process to promote excellence and innovation, and we commend those universities that already distribute their formula funds in this manner.

> **Recommendation 1a:** PCAST recommends that the focus of USDA research funding shift toward competitive grants, gradually rebalancing the research portfolio for intramural funding and funding for land grant institutions to incorporate incentives for innovation consistent with other research agencies across the Federal Government.

A healthy innovation ecosystem for agricultural research also requires a greater commitment to basic science relevant to the challenges to agriculture described previously, including plant, animal, and microbial biology; climatology; and health and nutritional research.

> **Recommendation 1b:** PCAST recommends an increase in funding for basic science relevant to agriculture.

This investment should be made primarily through the NSF, with participation from USDA, as well as other Federal agencies such as NIH and DOE when appropriate. NSF has demonstrated a successful model of program management and peer review that balances pure, merit-based review with additional considerations such as geographic distribution and historical legacies. We recommend a new investment, immediately doubling NSF research funding for basic agriculture-related science from a current level estimated at $120 million to $250 million for fiscal year 2014.[99] Over a 5year period, PCAST recommends that the overall Federal investment for basic science related to agriculture double as well. USDA participation in high-level program-development discussions within NSF and other agencies will help ensure coordination across the broad agricultural research effort, as was achieved in the National Plant Genome Research Program.[100, 101]

Recommendation 1c: PCAST also recommends an increase in competitively awarded funding within the USDA, raising the current level of funding for the Agriculture and Food Research Initiative (AFRI) from $265 million to $500 million (the original Congressional authorization was $700 million).

This new funding should be focused on the emerging challenges described in Chapter 2. The budget increase is consistent with the Administration's FY2013 Budget Request for AFRI of $325 million. New funding should be coordinated with NSF and other Federal science agencies, and representatives from NSF should participate in program design and funding decisions. In general, because of the long-term nature of the research to address these scientific challenges, awards should be of longer duration than the 2 to 3year awards that are common now.

Human Capital for U.S. Agriculture

To create a vibrant, innovative research enterprise, a primary concern is support for a well-trained workforce. We have heard from many of the experts who contributed to this report that the best students, particularly in the natural sciences, do not view agriculture, or agriculture-related research, as an attractive career option. At universities, relatively few graduate students enter into agricultural fields, and industry has difficultly recruiting the technical employees for its breeding and research programs, often turning to foreign students and workers, and then often employing them abroad. To meet the current and future challenges that agriculture will face, this situation must be reversed.

As successful as the land grant university system has been at training students for careers in agriculture, it is clear that some of our most talented students, especially first-generation college students, do not choose to attend agriculture schools or select traditional majors in agriculture or the ever-broadening set of scientific disciplines that contribute to agricultural research innovation. There are some notable exceptions in recent years, such as agronomy where starting salaries are setting new records as high commodity prices spur new interest, but for the most part, agriculture is facing a knowledge and workforce deficit.[102] Land grant universities and other research universities are facing a dual squeeze as both Federal and state budgets are cut at institutions that lack a history of major private-donor support. At the same time, some private universities, including those among the Nation's most elite schools, are renewing their focus on agriculture, water, energy, nutrition, and food systems because of student interest and the obvious strategic importance of these topics with respect to social and political stability, economic growth, and international development. Because of the historical exclusion of agricultural sciences from many of these colleges and universities, these students rarely receive the practical training provided in the more traditional agriculture schools.

First, we must recognize that the talent pipeline, especially for minority and first-generation college students from urban and rural backgrounds, begins well before college admission. A focus on secondary programs, such as the curriculum exemplified at the Chicago High School for Agricultural Science, hold tremendous potential to increase not only the number, but the diversity, of students entering baccalaureate programs, a requisite for the innovation we intend to spur. At the baccalaureate level, a comprehensive array of undergraduate programs relevant to agriculture and the food industry, as well as applied social and natural sciences, must remain strong and well-supported.

Consolidated centers of excellence with tuition-reciprocity agreements will allow land grant universities and other research universities to specialize and share expertise and programs. Such programs are now already commonplace as states share veterinary schools and develop specialized undergraduate programs such as the Midwest Poultry Consortium. Developing these centers of excellence will allow us on a national level to maintain broad access to diverse, key specialized agricultural programs, including agricultural education, without using limited resources to maintain a research or education operation in every historical location.

For the Nation to continue to provide the livestock and crops that feed the population, maintain the economic contributions of agriculture, and at the same time enhance nutrition, decrease the environmental footprint, and sustain yields in the face of an uncertain climate, careers in agriculture must be perceived as attractive to our brightest students. Further, the workforce must be well trained in traditional agriculture disciplines and emerging areas of biology, engineering, chemistry, computer science, and other mathematics-based sciences. While there are scattered success stories, for the most part, agriculture has not been able to attract sufficient numbers of students to careers in agriculture. In fact, a large proportion of the current students in graduate programs in agriculture are foreign.[103] The United States needs a national strategy for training the agricultural workforce to produce the innovations, technology, and products for the future. More important, strategic investment can be leveraged to provide attractive career options in both the public and private sectors for our brightest students, who now flock to careers in medicine, law, and business.

Recommendation 2: PCAST recommends that the USDA, in collaboration with NSF, expand a national competitive fellowship program for graduate students and post-doctoral researchers.

Graduate fellowship programs have a long history of attracting students to new fields, from plant sciences to nuclear physics and bioinformatics. Providing a strong, well-funded fellowship program will serve as a conduit for new scientists entering the field of agriculture. As with graduate fellowships, expansion of a postdoctoral fellowship program, perhaps modeled after the NSF Science, Engineering and Education for Sustainability fellowships, will attract the brightest students from diverse fields of basic science into the field of agriculture to pursue research on the emerging challenges described earlier. To leverage the substantial Federal investment in basic research across the natural sciences, and to specifically enhance participation in agriculture-related disciplines, these fellowship programs must be open to students and researchers at all types of research institutions—public, private, and land grant—and the fellowship programs must not be used simply to fund additional work within the large intramural program. The program including graduate and postdoctoral fellowships should be initially established at a level of $180 million per year with 5year funding.

In addition to offering graduate and postdoctoral fellowships, the USDA, in partnership with NSF, should continue to develop programs to attract young students to agricultural careers. Agriculture is invisible to the majority of high school students and is often not considered a desirable career option. Programs that introduce students to agriculture, such as the Chicago High School for Agricultural Sciences are in contrast to this trend. These programs not only expose students to the field of agriculture, but they have proven records of graduating students who then go on to choose careers in the field. The USDA should enhance its support for such programs to stimulate agriculture as a career option for students at the secondary school level.

Agricultural Research Infrastructure

Historically, agricultural research infrastructure has been developed through partnerships between the USDA Agricultural Research Service and a distributed state-administered system of research farms, typically tied to and maintained by the SAES, that receive both Federal and state support. In recent years, both Federal and state support has declined for the Agricultural Experiment Stations, sometimes markedly. The loss of Federal earmarks or special grants in FY2011 had an especially negative effect on this infrastructure. The USDA ARS capital budgeting process is particularly antiquated, reflecting political strategies that might have been justified in times of plenty, but that in today's budget environment can produce decades of delay in the construction of key facilities.

It is critically important looking forward to outline, and assign priorities within, the necessary national agricultural research infrastructure needed for a comprehensive and modern agricultural innovation ecosystem. As is the case with educational programs, each state can no longer afford to maintain the full spectrum of research facilities for all crops and livestock of current and future interest in that state. A modern system would build on the specialization that is already evident in some parts of the agricultural and food research system, for example, in animal health where a world class facility in Ames, IA, was dedicated in 2010, or in the USDA Federal nutrition laboratories located at Tufts University, Emory University, University of California at Davis, and ARS Beltsville, Maryland.

As an alternative to the current system, which is inadequate and falling further behind each year due to limited funds and conflicting priorities, a clear national plan for infrastructure should be developed. Such a plan should have mechanisms to ensure that existing strengths, now distributed across the decentralized ARS and state systems, are leveraged into consortia that in turn build on private sector investments. Funding for critical facilities should be awarded through a competitive process that balances local and regional needs with national priorities. This has already been a very successful strategy in a number of critical areas of agricultural research, such as cattle genetics, but there is great opportunity looking forward to expand this approach to support adequate infrastructure for both research and training.

Recommendation 3: PCAST recommends that the USDA expand its program of competitive awards for new infrastructure investments in agricultural research, with an emphasis on specialization and consolidation to avoid redundancies.

We recommend that the USDA convene a committee that evaluates the current infrastructure needs and sets priorities based on the research challenges described above. Awards should be made on a competitive basis, which will encourage all universities and the ARS to leverage their existing intellectual and capital resources to improve the overall quality of their programs.

Public-Private Partnerships

The private sector must be a major participant in the national research enterprise. Agricultural businesses bring a strong tradition of moving research results to product development and to commercial sales. Although many of the most critical new challenges in agriculture do not have an immediate impact on commercial interests, new product development and application will depend directly on developing solutions to the challenges and problems confronting agriculture.

There is no overarching structure in the United States that supports sustained, interactive research between public and private scientists interested in agricultural challenges as there is in fields such as nanotechnology and biofuels. Such a structure needs to be built and will likely require a clear definition of research responsibility for each aspect of the research, from conception to handling of data and publication. An open environment of investigation and publication is usually most beneficial, although the need to protect proprietary aspects of research must be considered for businesses to participate fully. The ultimate goal, however, needs to be a new structure for sustained close cooperation among all participants to expedite research. Such efforts should not be in areas where private research is already active as the issues relate directly to product development. New private-public partnerships should be created around those emerging challenges that do have some commercial interest, but that cannot be easily monetized in the short term.

Recommendation 4: PCAST recommends an initial new Federal investment of $150 million per year to create six institutes at a funding level of $25 million per year for no less than 5 years. Administration of this new program should be done by USDA, but closely coordinated with other Federal science agencies including NSF, DOE, and NIH, and representatives of these agencies should participate in the planning and funding decisions.

The USDA can begin immediately to invest in research toward meeting the challenges described earlier by establishing new innovation institutes supported by public-private partnerships, focused on addressing the specific challenges to agriculture. USDA could model these institutes after the bioenergy institutes established by DOE and BP or after the energy hubs and energy frontier research centers established by DOE. Consortia of private companies, universities, and researchers from the ARS could compete for large grants with support guaranteed for no less than 5 years to allow for bold new research investments. The research focus of each innovation institute must be on problems in the public domain, but where private sector participation can be important in advancing the research goals and also deploying the research outcomes.

Technology Development and Deployment

Scientific and technological discovery leads to new technologies, which themselves need to be developed and deployed. Venture capital firms have played a significant role in other technology developments (energy, IT, and pharmaceuticals) and have been increasing their investments in the life sciences area over the past two decades. In the agricultural sector, large companies play a dominant role, as there are significant hurdles that limit the ability of venture capital to seed startup companies. One obstacle that was repeatedly identified by members of the agricultural research community consulted for this report is the complex regulatory environment that delays or prevents new intellectual capital from being developed and translated into commercial products. A full discussion of regulatory policy for agriculture is beyond the scope of this report; however, creating a transparent regulatory system that protects public health and the environment, while encouraging commercialization of new products, is an essential component of the innovation ecosystem described herein.

Recommendation 5: PCAST recommends that the President request an internal review of Federal regulatory policy for agriculture to promote regulatory clarity, consistent with Executive Order 13563,[104] as well as with the Presidential Memorandum[105] on technology transfer from the national laboratories to the marketplace.

Furthermore, technology transfer from lab bench to marketplace has historically been underdeveloped within the agricultural research enterprise. This is due in part to the the proud history of the extension service and direct communication with farmers and producers, which often directly transferred knowledge rather than commercially developed it. However, as science and technology advance and become ever more sophisticated, and as taxpayers and policymakers seek to understand the commercial outcomes of research, it is critical that land grant university and ARS leadership continue to develop and strengthen their technology transfer efforts and thus bear the fruit of their research labors.

Implementation and Planning

This committee should include representatives from government agencies, including the White House Office of Science and Technology Policy (OSTP), USDA, NSF, NIH, and DOE; members of the university research community, including public, private, and land grant universities; as well as members from farmer and commodity groups, industry, and the venture capital community. The committee should develop an implementation strategy that guides the Federal investment in research, that develops enhanced competitive grants programs, and that coordinates the development of educational programs. The committee should be charged with designing new competitive programs to address the challenges described in this report, including the investments in education, infrastructure, and innovation institutes. This committee will ensure that the new Federal investment is available for open competition and is focused on those challenges having strong justification for public support. The committee should also undertake an assessment of the research portfolio to determine whether some current research investments are redundant with private-sector activities.

> **Recommendation 6:** PCAST recommends that the President establish an implementation committee to act on these recommendations.

Finally, as outlined in Chapters 1 and 2, agricultural research and related science and technology undergird the strength and long-term resiliency of the overall U.S. agricultural enterprise. The Federal investment in agricultural research should be administered to continually adjust priorities and focus on the most pressing challenges in as agile a manner as possible, while advocating strongly for a strong and broad basic science platform to provide future breakthroughs. To supplement the work of the implementation committee described above, a group of independent scientists and technologists from industry, academia, and nonprofit organizations can provide an important perspective, advising the Federal Government on how to best address the challenges facing agriculture and encouraging the USDA to pursue new opportunities and directions through time. PCAST recommends that a permanent, independent science advisory committee be formed to directly advise the Chief Scientist of the USDA. The advisory committee should provide an independent, science-based lens on the overall strength and direction of the agricultural research enterprise as governed by the USDA, as well as coordination with other U.S. research agencies, including the new investment through NSF described above.

Summary

PCAST believes that meeting the impending challenges to agriculture is an achievable goal. The United States has a university system that remains the envy of other nations and produces brilliant, innovative minds that have led the country to its preeminence across the various fields of science and technology. We have a vibrant private industry that has transformed the production of some crops and livestock by the application of new technology, improved biological systems, and frontier engineering. The U.S. venture capital and investment community has the experience and record for fostering technology transfer. And the United States, through the USDA, has an agricultural extension agency and enterprise that has been, and could continue to be, the model for the rest of the world. The raw material for a strong, vibrant innovation ecosystem is there. We need to develop a national strategy and a national will for bringing these components together, supporting the research enterprise where needed, and raising the profile of agriculture in rural and urban environments so that the best minds are attracted to careers in agricultural research. We believe that a new public investment in an agricultural research program designed for the challenges of the 21st century will not only enrich the U.S. heartland and communities everywhere, but will help the United States lead the world to a safer, healthier, more prosperous future.

APPENDIX A. METHODOLOGY

Commodity Groups

Checkoff groups were surveyed between May and July of 2012 by Don Latham via personal communication. Checkoffs were asked to report their total agriculture research funding and, where possible, to categorize their research expenditures. Pork, dairy, and lamb did not report data, and so estimates were made based on available information. Hass avocado, mango, and watermelon also did not report data, but given their relatively small size, data were not estimated. Only scientific research activities were included in data totals and were aggregated into the four categories in Table 1. Purchasing data, education, consumer behavior, retail management, outreach, and retail best practices research categories were excluded from the total.

Current Research Information System (CRIS)

USDA National Institute of Food and Agriculture (NIFA) tracks publicly funded USDA, state agriculture, and forestry research system activities through CRIS. CRIS reports ongoing agricultural, food science, human nutrition, and forestry research, education, and extension activities. Representatives of land grant universities and SAESs report funding expenditures on USDA-funded projects into CRIS. CRIS topical R&D breakdowns provide three taxonomies for agricultural R&D: knowledge area, subject of investigation, and field of science.[106] This allows organizations to report R&D funding by agriculture-relevant topics.

CRIS data have several limitations. First, CRIS covers R&D projects performed by land grant universities and SAESs that have been funded with at least some USDA funding. Therefore, CRIS data are subject to underreporting of non-land-grant-university performers and agriculture-related R&D wholly funded by non-USDA agencies, industry, commodity groups, and foundations. Second, CRIS' scope was designed to classify research performed by USDA and land grant universities. Thus, CRIS may undercount public R&D expenditures by excluding basic plant or animal science expenditures relevant to agriculture. Third, CRIS data do not add up to NIFA budget totals, as CRIS gathers data on research expenditures, which do not include some extension and education activities. Fourth, CRIS data are timelagged from budget data due to the time it takes for USDA R&D funds to be disbursed, allocated to projects, and reported back to CRIS.

National Center for Science and Engineering Statistics (NCSES)—Survey of Federal Funds for R&D

NCSES administers the Survey of Federal Funds for R&D annually. In this survey, Federal agencies report information regarding U.S. Federal funding for R&D, including breakdowns by agency and field. NCSES survey data for USDA do not capture certain expenditures considered to be R&D by this report's broad definition, including agricultural extension, some educational activities, and agricultural statistics.

National Center for Science and Engineering Statistics—Survey of R&D Expenditures at Universities and Colleges

The NCSES administered the "Survey of R&D Expenditures at Universities and Colleges"each year from 1972 to 2009.[107] This survey examines most university[108] R&D funding by academic field and the source of these funds. Given that its scope covers all types of R&D rather than only that funded or co-funded by USDA, it provides a broader array of biological and life science R&D funding data than CRIS.[109]

The primary limitation of data from the NCSES "Survey of R&D Expenditures at Universities and Colleges" is that the survey does not break out agricultural or biological R&D into smaller categories or provide taxonomies for agricultural R&D. It is thus impossible to know how much of the "biological" discipline may be relevant to agriculture, as basic plant and animal biology are grouped together with all other types of biological science.[110]

USDA Economic Research Service

In 2011, USDA Economic Research Service published a report on private R&D investments for agricultural inputs, food processing, and biofuels by Fuglie et. al called "Research Investments and Market Structure in the Food Processing, Agricultural Input, and Biofuel Industries Worldwide."[111] To estimate U.S. agricultural inputs R&D expenditures, Fuglie et. al used information from firms' annual financial reports, industry associations, consulting services, and interviews. To estimate U.S. food manufacturing R&D expenditures, they used country-level estimates from the Organisation for Economic Cooperation and Development's Analytical Business Enterprise Research and Development database. To estimate biofuel R&D estimates, they gathered information on sectors doing R&D on biofuel feedstocks and manufacturing.[112] The Science and Technology Policy Institution (STPI) estimated U.S. foundations agricultural R&D expenditures from a personal conversation with Fuglie.

Data from "Research Investments and Market Structure in the Food Processing, Agricultural Input, and Biofuel Industries Worldwide" have several limitations for this study. First, U.S. agricultural inputs estimates from Fuglie et. al count only R&D performed by companies incorporated in the United States, assuming that the R&D performed by U.S. companies outside the country is balanced by companies incorporated outside the United States that do R&D inside the country. Second, while the data cover most traditional agricultural companies, they do not cover all areas of agricultural research captured in CRIS or NCSES, such as paper companies that perform forestry R&D and biopharmaceutical companies doing basic plant R&D. Third, agricultural inputs and food manufacturing R&D estimates may include biofuel R&D expenditures, although this report attempts to remove such double counting. Finally, while the report covers data for some sectors up to 2009, data were not available for all sectors to 2009. Because this report focused on recent funding levels, certain sectors' funding levels were taken from the most recent year available (2006–2008).

USDA Research, Education, and Economics (REE) Budget

The USDA REE budget represents the portion of the USDA annual budget dedicated to R&D, education including student fellowships, extension activities, and economic and statistical analysis and research. REE budget numbers provide the broadest potential definition for R&D from USDA, as they include activities such as extension services, agricultural statistics, and education programs that are not always counted as research.[113] The REE portion of the USDA budget does not include Forest Service R&D, which was added from the broader USDA budget.

Limitations to Combined Data Sources

Because no one data source provides a comprehensive view of agricultural R&D activities in the U.S., it was necessary to combine four sources—commodity groups, CRIS, USDA ERS, and USDA REE budget—to achieve the complete picture of U.S. agricultural R&D expenditures represented in Figures 1 and 2. Due to the limitations of each source as listed before, combining data sources produces uncertainty in the estimates. Further uncertainties in Figures 1 and 2 are due to different reporting periods for different sources, including different fiscal years and the time lag described for CRIS data. While the previous descriptions of data sources highlight where undercounting of R&D expenditures might occur due to limited scope of performers or limited scope of field of research, combining data sources likely leads to double counting of agricultural R&D expenditures. For example, CRIS values for nonpublic expenditures at land grant universities and SAESs may duplicate values for the private sector and foundation R&D estimates. This report assumes that data from different sources are additive, with the exception of commodity groups, which are expected largely to fund research at land grant universities and be reported in CRIS.

APPENDIX B. ADDITIONAL EXPERTS PROVIDING INPUT

Jane Anders
Vice President, Research, Quality, and Innovation
ConAgra Foods

Mike Edgerton
Technology Lead for Corn Ethanol and Quality
Traits

David Battisti
Professor of Atmospheric Sciences
University of Washington
Roger Beachy
Professor of Biology
Washington University of St. Louis
Jim Birchler
Curators Professor of Biological Sciences
University of Missouri
Ed Buckler
Research Geneticist
USDA Agricultural Research Service
Adjunct Professor
Cornell University
Steven Burrill
Chief Executive Officer
Burrill & Company
Jim Carrington
President
Danforth Plant Science Center
R. James Cook
Professor Emeritus
Washington State University
Jack Dixon
Vice President and Chief Scientific Officer
Howard Hughes Medical Institute
Harvard University
Richard A. Dixon
Senior Vice President, Professor
Director of the Plant Biology Division
Samuel Roberts Noble Foundation
Richard Jackson
Professor and Chair, Environmental Health
Sciences
University of California Los Angeles
Scott Jackson
Professor, Center for Applied Genetic Technologies
University of Georgia
Steve Koonin
Former Undersecretary for Science
Department of Energy
Monsanto
Nina Fedoroff
Evan Pugh Professor
Penn State University
Distinguished Visiting Professor
King Abdullah University of Science & Technology
Robb Fraley
Executive Vice President and Chief Technology
Officer
Monsanto
Keith Fuglie
Chief of the Resource, Environmental, and Science Policy Branch
USDA Economic Research Service
Dan Glickman
Former Secretary of Agriculture
Former Congressman
Senior Fellow, Bipartisan Policy Center
Ray Goldberg
Professor, Agriculture and Business, Emeritus
Harvard Business School
M.R.C. Greenwood
President
University of Hawaii
Jerry Hjelle
Vice President, Regulatory Affairs
Monsanto
Peter Huybers
Assistant Professor
Department of Earth and Planetary Sciences
Harvard University
Michael Roberts
Associate Professor, Department of Agricultural
and Resource Economics
North Carolina State University
Pamela Ronald
Professor, Plant Pathology
University of California Davis
Director

Director, Center for Urban Science and Progress
New York University
Upmanu Lall
Professor of Engineering, Department of Earth and Environmental Engineering
Columbia University
Brian Larkins
Porterfield and Regents Professor, School of Plant Sciences, Molecular and Cellular Biology
University of Arizona
Steve Long
Professor, College of Agricultural, Consumer and Environmental Sciences
University of Illinois
Consuelo Madere
Vice President, Vegetable Seeds Division and Asia Commercial
Monsanto
Francis Pierce
Professor Emeritus
Washington State University
AgInfomatics, LLC
Charles Rice
Professor of Soil Microbiology
Kansas State University
Joint Bioenergy Institute
Robin Schoen
Director
Board on Agriculture and Natural Resources
National Resource Council
Norman R. Scott
Professor Emeritus
Cornell University
Chairman, Board on Agriculture and Natural Resources
National Resource Council
John Soper
Vice President, Crop Genetics Research & Development
Pioneer, DuPont
David Stern
Scientist
Boyce Thompson Institute
David Tilman
Professor and Chair, Department of Ecology
University of Minnesota
Walter Willett
Professor of Epidemiology and Nutrition
Harvard University
Catherine Woteki
Undersecretary for Research, Education, and Economics
Department of Agriculture
Roger Wyse
Managing Director
Burrill & Compan

APPENDIX C. ACKNOWLEDGMENTS

Joe Alper
Science Writer
Life Science and Nanotechnology Consulting
Wanda Archy
Chris Weber
IDA Science and Technology Policy Institute
Caren Wilcox
Special Assistant

Student Volunteer
Office of Science and Technology Policy
Kaitlin Bernell
Student Volunteer
Office of Science and Technology Policy
Ben Buchanan
Student Volunteer
Office of Science and Technology Policy
Genevieve Croft
Student Volunteer
Office of Science and Technology Policy
Knatokie Ford
AAAS Science and Technology Policy Fellow
Office of Science and Technology Policy
Paul Heisey
Agricultural Economist
USDA Economic Research Service
Kei Koizumi
Assistant Director for Federal R&D
Office of Science and Technology Policy
Falita Liles
Information Technology Specialist
USDA National Institute of Food and Agriculture
Brent Miller
IDA Science and Technology Policy Institute
Gina Walejko
IDA Science and Technology Policy Institute
USDA Office of the Undersecretary for Research,
Education, and Economics
Richmond Wong
Student Volunteer
Office of Science and Technology Policy

End Notes

[1] Details are outlined in Chapter 4. Briefly, $180 million for new graduate and post-doctoral fellowships; $235 million for new competitively funded research at USDA AFRI; $130 million for basic research at NSF; and $150 million for new public-private institutes.

[2] Economic Research Service, US Department of Agriculture. (2012). "Value of U.S. trade—agricultural, nonagricultural, and total—and trade balance, by calendar year." Accessed May 18, 2012 at www.ers.usda.gov datafatus.

[3] Public comments from USDA Secretary Tom Vilsack before PCAST, March 9, 2012. Accessed August 24, 2012 at www.tvworldwide.com/events/pcast/120309/globe_show/default_go_archive.cfm?gsid=1977&type=flv &test=0&live=0.

[4] National Research Council. (2009). *A New Biology for the 21st Century.* Accessed June 19, 2012 at www.ncbi.nlm.nih.gov/books/NBK 32509.

[5] Alston, J.M., M.A. Andersen, J.S. James, and P.G. Pardey. (2010). *Persistence Pays: U.S. Agricultural Productivity Growth and the Benefits from Public R&D Spending.* "Chapter 2 A Brief History of U.S. Agriculture." 504.

[6] Egli, D.B. (2008). "Comparison of Corn and Soybean Yields in the United States: Historical Trends and Future Prospects." *Agronomy Journal*, "Celebrate the Centennial" supplement. S7988.

[7] Evenson, R.E. and D. Gollin. (2003). "Assessing the impact of the Green Revolution, 19602000." *Science 300:* 758-762.

[8] Pretty, J. (2008). "Agricultural sustainability: concepts, principles and evidence." P*hilosophical Transactions of the Royal Society* B 363: 447-465.

[9] Congressional Research Service. (23 March 2012). "Agricultural Research, Education, and Extension: Issues and Background." Figure 6:8.

[10] NSF NCSES. Data from Federal Funds for Research and Development: Fiscal Years 2008–10. Accessed August 24, 2012 at www.nsf.gov/statistics/nsf12308/content.cfm?pub_id=4121&id=2.

[11] Note: For NSF and NIH, total extramural funding budget numbers were used as a proxy for competitive funding amounts because these two funding agencies award all of their extramural research grants through competitive, merit review. Similar estimations are not possible for DOD, DOE, or NASA, which are also agricultural research funders presented in Figure 3.

[12] USDA ERS. (2011). Global Public Agricultural Research Spending. Graph derived from OECD, Eurostat and ASTI. The graph indicates that China, India, and Brazil have made steady and continuous investments in agriculture research over the last decade in particular.

[13] Hodson, D. and E. Depauw. (2011). Use of GIS Applications to Combat the Threat of Emerging Virulent Wheat Stem Rust Races: 129157. In: Clay, S. (Ed.) GIS Applications in Agriculture 3: Invasive Species. CRC Press, Boca Raton, FL.

[14] Stokstad, E. (2012). "Dread Citrus Disease Turns Up in California, Texas." *Science* 336: 283284.

[15] National Research Council. (2010). "Strategic Planning for the Florida Citrus Industry: Addressing Citrus Greening." Accessed June 25, 2012 at www.nap.edu/catalog.php?record_id=12880.

[16] National Research Council. (2004). California Agricultural Research Priorities: Pierce's Disease. Accessed June 26, 2012 at www.nap.edu/catalog.php?record_id=11060.

[17] Environmental Protection Agency (EPA). Ozone layer protection—regulatory programs: "Phaseout of methyl bromide." Accessed August 24, 2012 at www.epa.gov/ozone/mbr.

[18] USDA ERS. "ERS investigates and quantifies water allocation, water conservation, and water management issues facing irrigated agriculture." Accessed May 21, 2012 at www.ers.usda.gov/briefing/wateruse.

[19] United Nations Food and Agriculture Organization (FAO). "Improving irrigated production." Accessed June 15, 2012 at www.fao.org/DOCREP/005/Y3918E/y3918e10.htm.

[20] Rabalais, N.N., R.E. Turner, and W.J. Wiseman. (2002). "Gulf of Mexico Hypoxia, a.k.a. 'The Dead Zone.'" *Annual Review of Ecology and Systematics* 33: 235-263.

[21] Marder, J. (2011). "Farm Runoff in Mississippi River Floodwater Fuels Dead Zone in Gulf." Accessed June 15, 2012 at www.pbs.org/newshour/rundown/2011/05/thegulfofmexicohas.html.

[22] Tilman, D. Global. (1999). "Environmental impacts of agricultural expansion: The need for sustainable and efficient practices." *Proceedings of the National Academy of Sciences* 96(11): 59956000.

[23] USDA ERS. "Land use statistics summary." Accessed May 18, 2012 at www.ers.usda.gov/Briefing/LandUse.

[24.] *Ibid.*

[25] President's Council of Advisors on Science and Technology (PCAST). (2011). *Report to the President on Sustaining Environmental Capital: Protecting Society and the Economy.* Accessed August 24, 2012 at www.WhiteHouse.gov/sites/default/files/microsites/ostp/pcast_sustaining_environmental_capital_report .pdf.

[26] Janzen, H. et al. (2011). "Global prospects rooted in the soil." *Soil Science Society of America Journal* 75: 1-8.

[27] EPA. (2012). Inventory of US Greenhouse Gas Emissions and Sinks: 1990-2010. Accessed May 18, 2012 at www.epa.gov/climatechange/emissions/usinventoryreport.html.

[28] Schlenker, W. and M. J. Roberts. (2009). "Nonlinear temperature effects indicate severe damages to U.S. crop yields under climate change." *Proceeding of the National Academy of Sciences* 106(37): 1559415598.

[29] Lobell, D.B., M. Bänziger, C. Magorokosho, and B. Vivek. (2011). "Nonlinear heat effects on African maize as evidenced by historical yield trials." *Nature Climate Change* 1: 42-45.

[30] Wegren, S.K. (2011). "Food security and Russia's 2010 drought." *Eurasian Geography and Economics* 52 (1): 140-156.

[31] Carroll, A.L., S.W. Taylor, J. Regniere, and L. Safranyik. (2003). "Effects of climate change on range expansion by the mountain pine beetle in British Columbia." 223-232 in Shore, T.L., J.E. Brooks, and J.E. Stone (eds.), Proceedings of the Symposium on Mountain Pine Beetle Symposium: Challenges and Solutions. Oct. 30-31, 2003, Kelowna, British Columbia. Information Rpt. BC-X-399. Natural Resources Canada, Canadian Forest Service, and the Pacific Forestry Centre, Victoria, BC.

[32] Battisti, D.S., J.J. Tewksbury, and C.A. Deutsch. The impact of global warming on global crop yields due to changes in pest pressure. American Geophysical Union, Fall Meeting 2011, abstract 3#U53F-04.

[33] Mote, P.W., A.F. Hamlet, M.P. Clark, and D.P. Lettenmaier. (2005). "Declining mountain snowpack in Western North America. Bull." *American Meteorological Society* 86(1):39–49.

[34] Westerling, A.L., H.G. Hidalgo, D.R. Cayan, and T.W. Swetnam. (2006). "Warming and Earlier Spring Increase Western U.S. Forest Wildfire Activity." *Science* 313: 940-943.

[35] USDA FAS. (2011). Foreign Agriculture Service Report: "U.S. on Track to become World's Largest Ethanol Exporter in 2011." Accessed August 24, 2012 at www.fas.usda.gov/info/IATR/072011_Ethanol_IATR.pdf; and Urbanchuk, J.M. (2012). Contribution of the ethanol industry to the economy of the United States. Remarks prepared for the Renewable Fuels Association. Accessed August 24, 2012 at ethanolrfa.3cdn.net/c0db7443e48926e95f_j7m6i6zi2.pdf.

[36] Kazi, F. K., et al. (2010). TechnoEconomic Analysis of Biochemical Scenarios for Production of Cellulosic Ethanol. Technical Report. NREL/TP6A246588. Accessed May 18, 2012 at www.nrel.gov/docs/fy10osti/46588.pdf.

[37] Centers for Disease Control (CDC). (2012). "Weight of the Nation." Accessed May 18, 2012 at www.cdc.gov/media/matte/2012/05_weight_of_nation.pdf.

[38] CDC Data. (2012). Accessed June 19, 2012 at www.cdc.gov/diabetes/statistics/prev/national/figbyage.htm. Diabetes increased between 1980 and 2010 200 percent (from 0.6 percent to 1.8 percent) for those aged 0–44 years, 124 percent (from 5.5 percent to 12.3 percent) for those aged 45–64 years, 127 percent (9.1 percent to 20.7 percent) for those aged 65–74 years, and 126 percent (8.9 percent to 20.1 percent) for those aged 75 years and older

[39] CDC. 2012. "Weight of the Nation." Accessed May 18, 2012 at www.cdc.gov/media/matte/2012/05_weight_of_nation.pdf.

[40] USDA ERS. "Are Healthy Foods Really More Expensive? It Depends on How You Measure the Price." Accessed August 24, 2012 at www.ers.usda.gov/publications/eib-economic-information-bulletin/eib96.aspx.

[41] Lustig, R.H., L.A. Schmidt, and C.D. Brindis. (2012). "The toxic truth about sugar." *Nature* 482: 27029.

[42] U.S. Census Bureau. (1995). Statistical Abstract of the United States 1995: 664. Accessed July 30, 2012 at www.census.gov/prod/1/gen/95statab/agricult.pdf.

[43] World Bank statistics. U.S. Census Bureau. (2012). Statistical Abstract of the United States2012: 536. Accessed July 30, 2012 at www.census.gov/prod/2011pubs/12statab/agricult.pdf.

[44] This example emphasizes the crucial need to maintain genetic diversity in all food crops to prevent putting the Nation's food supply at risk. "Blight in the corn belt." Accessed August 23, 2012 at www2.nau.edu/~bio372c/class/sex/cornbl.htm.

[45] USDA ERS. International Food Security Assessment: 2012-22. Accessed August 8, 2012 at www.ers.usda.gov/publications/gfa-food-security-assessment-situation-and-outlook/gfa23.aspx.

[46] Tilman, et al. (2011). "Global food demand and the sustainable intensification of agriculture."*Proceedings of the National Academy of Sciences.* Vol. 108.

[47] Helsey, P., S. L. Wang, and K. Fuglie. (2011). Public agricultural research spending and future U.S. agricultural productivity growth: scenarios for 20102050. USDA ERS Economic Brief 17. Accessed August 24, 2011 at www.ers.usda.gov/media/118663/eb17.pdf.

[48] Rosenberg, D. (2011). "Food and the Arab Spring." Accessed May 18, 2012 at www.gloriacenter.org/2011/10/food-and-the-arab-spring/#_edn46.

[49] Some early reporting on the protests that became known as the Arab Spring: Cha, A. E. (14 January 2011). "Spike in Global Food Prices Contributes to Tunisian Violence." *The Washington Post.*; McDevitt, J. (15 January 2011). "Jordanians Protest Against Soaring Food Prices." *The Guardian.*; Geewax, M. (30 January 2011). "Rising Food Prices Can Topple Governments, Too." National Public Radio.; Rubin, J.. (9 February 2011). "Food: What's Really Behind the Unrest in Egypt." *Toronto Globe & Mail.*

[50] President Obama's remarks to the G-8, Chicago, IL. Accessed June 11, 2012 at www.whitehouse.gov/thepress-office/2012/05/18/remarks-president-symposium-global-agriculture-and-food-security.

[51] United Nations Food and Agriculture Organization (FAO). "Dimensions of need – An Atlas of food and agriculture." (1995). Chapter on Staple foods: What do people eat?" Accessed May 19, 2012 at www.fao.org/docrep/U8480E/U8480E00.htm.

[52] Food and Agriculture Organization. (2012). FAOSTAT. Accessed August 22, 2012 at faostat.fao.org/site/567/default.aspx#ancor. As an example, the 2010 maize yield for North America is ~96 Hg/ Ha, while the yield for middle Africa is ~10 Hg/Ha.

[53] Fuglie, K.O., and P. W. Heisey. 2007. "Economic Returns to Public Agricultural Research." United States Department of Agriculture Economic Research Service. *Economic Brief Number 10.*

[54] Note: The terms "R&D" and "research" are used interchangeably throughout the report, but it is important to note that different sectors have unique R&D portfolios. USDA R&D represents roughly 90 percent research and 10 percent development, whereas, most accounts indicate that the private sector agricultural research portfolios are strongly weighted towards development. (Office of Management and Budget, *Budget of the US Government FY 2013*, Analytical Perspectives Table 22-1.)In Figures 1 and 2, R&D amounts represent specific data that cover both research and development according to standard measures. For clarification on this and related methodology, see Appendix A.

[55] Fuglie, K. O., et al. (December 2011). "Research Investments and Market Structure in the Food Processing, Agricultural Input, and Biofuel Industries Worldwide," U.S. Department of Agriculture, Economic Research Report Number 130.

[56] Definition adopted from the Current Research Information Systems Manual of Classification for Agricultural and Forestry Research, Education, and Extension. Accessed August 8, 2012 at www.ers.usda.gov/data-products/agricultural-research-funding-in-the-public-and-private-sectors/definitions-and-related-resources.aspx.

[57] An LGU is a higher education institution that has been designated by its State legislature or Congress to receive unique Federal support based on the Morrill Acts of 1862 and 1890. A map of LandGrant Colleges and Universities can be accessed at www. csrees.usda.gov/qlinks/partners/partners_map.pdf (accessed August 22, 2012). State agricultural experiment stations were originally created in each State to link research with the educational mission of the LGUs, as a result of the Hatch Act of 1887. They are generally affiliated with LGUs.

[58] Total amounts of Federal funding for agricultural research depend on the definition of "agricultural research" and the data source. For more details of estimates and methods used, see Figure 3 legend and Appendix A.

[59] A small proportion of the funds accounted for in CRIS ($58M of $5.1B in FY09) is identified as "other cooperating institutions" that represents performers of National Institute of Food and Agriculture-funded research outside of LGUs and SAESs.

[60] Many R&D portfolio comparisons, including the NCSES data presented in Figure 4, do not include extension funding within the R&D category. However, because extension is listed in Figures 1-2, it is important to note that the inclusion of extension in the total USDA R&D budget would bring the total extramural proportion to approximately 43 percent.

[61] The data shown in Figure 4 do not include Extension Services. If using Figure 1 data are used and including Extension Services, this number drops to 55 percent.

[62] Note: Some other Federal science-based agencies that also have large fraction of intramural funding are not listed here, including the Department of Interior (which includes the U.S. Geological Survey) and the Department of Commerce (which includes the National Oceanic and Atmospheric Administration). We exclude these agencies because they do not have a primary responsibility for Federal support of extramural research (i.e., these areas are also funded by NSF, DOE and NASA).

[63] Fuglie, K., et al. (1996). "Agricultural Research and Development: Public and Private Investments Under Alternative Markets and Institutions." USDA ERS Agricultural Economic Report No. (AER-735).

[64] McDougal, P. (2012). "Trends in Industry Research and Development." *Agrifutura* newsletter 150.

[65] Comparatively, the U.S. pharmaceutical industry spent about $67 billion, approximately 17 percent of its total sales, on R&D in 2011. Source: Research!America biomedical research industry survey. Accessed May 30, 2012 at www.researchamerica.org/uploads/healthdollar10.pdf. Industry-reported data from the Pharmaceutical Research and Manufacturing Association (PhRMA). Accessed May 30, 2012 at www.phrma.org/sites/default/files/159/phrma_profile_2011_final.pdf.

[66] Note: We have chosen to include the total agricultural R&D spending by the major multinational agricultural companies including those based in and outside of the United States. Most of the R&D conducted by these companies benefits U.S. farmers, and the United States is the largest single market for the products developed by these companies.

[67] Fuglie et al. (2011). "Research Investments and Market Structure in the Food Processing, Agricultural Input, and Biofuel Industries Worldwide," USDA, Economic Research Service Report No 130, Table 1.7.

[68] *Ibid.*

[69] Data provided by Dan Broderick of the National Venture Capital Association in May 2012, based on a Thomson Reuters investment search of currently or formerly private-equity/venture-capital backed portfolios, or portfolios of unknown status except real estate properties in the following industries: Agriculture, Forestry, Fishing, Animal Husbandry, etc., Agriculture related, Animal husbandry, Other Agriculture, Forestry, Fishing, Water Treatment Equipment & Waste Disposal Systems, and Other.

[70] Data from Dow Jones Venture Source database query of agriculture and forestry sector deals. Deal & Dollar Raised for US Venture-Backed Companies, by Industry (20052011). Accessed June 26, 2012 at www. venturesource.com.

[71] Interview with Ray Rothrock, Chairman of the National Venture Capital Association, in May 2012 who estimates that 32 percent of energy venture deals are directly spun from university labs.

[72] Data from Dow Jones Venture Source database query of energy sector deals. Accessed on June 26, 2012. Deal & Dollar Raised for US Venture-Backed Companies, by Industry (20052011) at www.venturesource.com.

[73] Ibid.

[74] Personal communication from the checkoff programs to Don Latham (Working Group member) upon request. Data compiled May 2012. Only research dollars in four broad categories were included in totals: Nutrition & Health, Production, New Uses, and Food Safety & Disease.

[75] "Overview of the Leveraging of Checkoff Funds" by John Becherer, February 22, 2012, United Soybean Board Board Meeting.

[76] Personal communication from the checkoff programs to Don Latham and Molly Jahn (Working Group members) upon request. Data compiled May 2012. Only research dollars in four broad categories were included in totals: Nutrition & Health, Production, New Uses, and Food Safety & Disease. See Appendix A for more detail.

[77] Derived from National Center for Science and Engineering Statistics, Academic Research and Development Expenditures: Fiscal Year 2009, Table 50.Accessed August 24, 2012 at www.nsf.gov/statistics/nsf11313/content.cfm ?pub_id=4065&id=2.

[78] National Research Council. 1996. Committee on the Future of the Colleges of Agriculture in the Land Grant University System. Report on Colleges of Agriculture at the Land Grant Universities: Public Service and Public Policy; Alston and Pardey. 1996. Making Science Pay: the Economics of Agricultural R&D Policy.

[79] NIH Research Portfolio Online Reporting Tools (RePORT), 2012 data. Accessed August 23, 2012 at http://report.nih.gov/award/organizations.cfm.

[80] CHSAS website. Accessed October 23, 2012 at www.chicagoagr.org/about/ principals_message.jsp

[81] A Suntimes publication reported that "demand for the school is soaring as more than 3,000 applications were submitted for only 150 freshman spots." Accessed October 23, 2012 at southtownstar.suntimes.com/business/10469839420/quinn-at-ag-school-jobs-follow-brain-power.html.

[82] NIFA was created by Congress in the Food, Conservation, and Energy Act of 2008. NIFA replaced the former Cooperative State Research, Education, and Extension Service (CSREES), which had been in existence since 1994.

[83] Agricultural Research Service, USDA. About ARS. Accessed August 23, 2012 at www.ars.usda.gov/aboutus/about-us.htm.

[84] Agricultural Research Service, USDA. Research. Accessed August 23, 2012 at www.ars.usda.gov/research/programs.htm.

[85] Economic Research Service, USDA. About ERS. Accessed August 23, 2012 at www.ers. usda.gov/abouters.aspx.

[86] United States Forest Service. "US Forest Service Research & Development." Accessed August 23, 2012 at www.fs.fed.us/research.

[87] National Agricultural Statistics Service, USDA. "About NASS." Accessed August 23, 2012 at www.nass.usda.gov/About_NASS/index.asp.

[88] U.S. Department of Agriculture, National Institute of Food and Agriculture. "About Us Extension." Accessed October 10, 2012 at http://www.csrees.usda.gov/qlinks/extension. html.

[89] Council of Economic Advisers. (June 2012). "Strengthening rural communities: Lessons from a growing farm economy." Figure 5: Public and Private U.S. Agricultural R&D Spending,

19712009. Total public funding has hovered between approximately $4.5B and $5B for the last two decades.

[90] Congressional Research Service. (23 March 2012). "Agricultural Research, Education, and Extension: Issues and Background." Figure 6:8.

[91] See Chapter 3 for an indepth discussion of these two findings.

[92] Council of Economic Advisers. (June 2012). "Strengthening rural communities: Lessons from a growing farm economy."

[93] National Research Council. (2009). "A New Biology for the 21st Century." Accessed June 19, 2012 at www.ncbi.nlm.nih.gov/books/NBK 32509.

[94] Task Force of the United States Department of Agriculture 2004. National Institute for Food and Agriculture: A Proposal. Accessed July 10, 2012 at www.ars.usda.gov/sp2userfiles/place/00000000/national.doc on July 10.

[95] National Research Council. (1972). Report of the Committee on Research Advisory to the U.S. Department of Agriculture. National Technical Information Service, Springfield, VA:393-394.

[96] National Research Council. (1989). Investing in Research: A Proposal to Strengthen the Agricultural, Food, and Environmental System. National Academy Press, Washington, DC. Accessed July 11, 2012 at www.nap.edu/catalog/1397.html on July 11, 2012.

[97] National Research Council. (2000). National Research Initiative: A Vital Grants Program in Food, Fiber, and Natural-Resources Research. National Academy Press, Washington, DC. Accessed July 11, 2012 at www.nap.edu/catalog/9844.html on July 11, 2012.

[98] National Research Council. (2003). Frontiers in Agriculture Research: Food, Health, Environment, and Communities. National Academies Press, Washington, DC:7.

[99] See Figure 3, third bar, "Upper CRIS," NSF investment is estimated at $121 million.

[100] 110th Congress, Section 7406 of the Food, Conservation, and Energy Act of 2008 (Pub. L. 110-246) [H.R. 2419].

[101] The President's FY 2013 Budget request included an increase for AFRI from $250 million to $323 million and indicates the Administration's commitment to competitive funding for agricultural research.

[102] According to data from the Crop Science Society of America, the scientific professional society for U.S. crop scientists, the number of crop scientists has significantly declined in the past two decades. Specifically, total membership of the Society's two major disciplines, breeders and physiologists, has decreased 56 percent from 1990 to 2010, the majority from the public sector.

[103] NSF NCSES data. Accessed June 15, 2012 at www.nsf.gov/statistics/ nsf12300/pdf/nsf12300.pdf. Data from the report "Graduate Students and Postdoctorates in Science and Engineering: Fall 2009" shows that temporary visa holders make up: 31 percent of fulltime graduate students in all agriculture related areas; 26 percent of fulltime agricultural science graduate students, 55 percent of fulltime agricultural engineering graduate students; 51 percent of full-time agricultural economics graduate students; 21 percent of all agricultural science graduate students; and 52 percent of all agricultural engineering graduate students.

[104] Executive Order 13563—Improving regulation and regulatory review. (January 18, 2011)

[105] Presidential Memorandum—Accelerating technology transfer and commercialization of Federal research in support of high-growth businesses. (October 28, 2011).

[106] The CRIS subject of investigation, knowledge area, and field of science classification schemes can be found at cris.nifa.usda.gov/manual.html.

[107] More recent data are collected by the Higher Education R&D Survey, successor to the Survey of R&DExpenditures at Universities and Colleges.

[108] The Survey of R&D Expenditures at Universities and Colleges surveys all institutions that budget at least $150,000 to science and engineering R&D.

[109] NCSES classifications can be found on the survey instrument at www.nsf.gov/statistics/nsf 11313/.

[110] *Ibid.*

[111] Fuglie, et al. (2011). "Research Investments and Market Structure in the Food Processing, Agricultural Input, and Biofuel Industries Worldwide." USDA ERS Economic Information Bulletin No. (EIB90). Accessed August 24, 2012 at http://www.ers.usda.gov/media/193646/eib90_1_.pdf.

[112] The U.S. share of biofuel R&D was based on this information but supplemented by a personal conversation with K. Fuglie.

[113] The 2009 REE budget (including the Forest Service) reports $3020 million in R&D funding for USDA, while the NCSES Survey of Federal Funds reports only $2265 million. This difference is primarily due to lower NCSES values for NASS and NIFA, likely caused by an undercounting of education activities, extension services, and statistical analysis.

In: U.S. Agricultural Preparedness ...
Editor: Warren C. Young

ISBN: 978-1-62618-437-4

Chapter 2

AGRICULTURAL RESEARCH, EDUCATION, AND EXTENSION: ISSUES AND BACKGROUND*

Dennis A. Shields

SUMMARY

The U.S. Department of Agriculture (USDA) Research, Education, and Economics (REE) mission area has the primary federal responsibility of advancing scientific knowledge for agriculture through research, education, and extension. USDA REE responsibilities are carried out by four agencies: the Agricultural Research Service (ARS), the National Institute of Food and Agriculture (NIFA), the Economic Research Service (ERS), and the National Agricultural Statistics Service (NASS). The USDA administers extramural federal appropriations to states and local partners primarily through three funding mechanisms: formula funds, competitive grants, and non-competitive grant programs.

The FY2012 Agriculture Appropriations Act (P.L. 112-55, H.R. 2112) contained $2.53 billion in discretionary funds for USDA agricultural research, education, and extension programs. While inflation-adjusted public spending for agricultural research grew steadily from the 1950s to the 1970s, it has

* This is an edited, reformatted and augmented version of the Congressional Research Service Publication, CRS Report for Congress R40819, dated March 23, 2012.

remained relatively flat since the 1970s (with a few exceptions), and growth in funding has lagged behind that of other national science agencies.

The enacted 2008 farm bill (P.L. 110-246) directed USDA to reorganize the REE mission area. The farm bill created a new entity called the National Institute for Food and Agriculture (NIFA), which assumed all programs and authorities from the Cooperative State Research, Education, and Extension Service (CSREES). A new competitive grants program for basic and applied research, called the Agriculture and Food Research Initiative (AFRI), was also established by the 2008 farm bill and is administered by NIFA. The farm bill also extends and expands mandatory and discretionary funding for specialty and organic crops research, bioenergy programs, and pollinator protection programs, among others. Several mandatory programs (research and otherwise) that were authorized in the 2008 farm bill do not have a budget baseline that extends beyond the end of the 2008 farm bill (September 30, 2012). If policymakers want to continue these programs in the next farm bill, they will need to pay for them with other offsets.

Debates over the direction of public agricultural research and the nature of its funding mechanism continue. Ongoing issues include the need, if any, for new federal funding to support agricultural research, education, and extension activities, and the implications of allocating federal funds via formula funds versus competitive grants. In addition, factors including the growing importance of specialty crops, international trade negotiations, and a renewed interest in international agricultural development have many groups believing that Congress needs to increase support of U.S. agriculture through expanded research, education, and extension programs, whereas others believe that the private sector, not taxpayer dollars, should be used to support these activities.

While the 2008 farm bill provided a significant funding boost for agricultural research, relatively speaking, it is unclear whether budgetary resources and political will can sustain funding for agricultural research and related activities.

INTRODUCTION

Public investment in agricultural research has been linked to productivity gains, and subsequently to increased agricultural and economic growth.[1] Studies consistently find high social rates of return on average from public agricultural research, widely reported to be in the range of 20%-60% annually.[2]

Advances in agricultural research, education, and extension were critical to the huge agricultural productivity gains seen in the United States after World War II (**Figure 1**).[3] Agricultural productivity grew on average by about 2%-3% percent annually during the 1950s through the 1980s, but growth has slowed in recent decades (**Figure 2**). Advances in the basic and applied agricultural sciences, such as disease-resistant crop varieties, efficient irrigation practices, and improved marketing systems, are considered fundamental to achievements in agricultural yields, increases in farm sector profitability, higher competitiveness in international agricultural trade, and improvements in nutrition and human health.

Some see a need for increased public investment in agricultural research in order to maintain high productivity growth. Congress determines the level of federal funding for agricultural research, education, and extension and must prioritize these activities in relation to other federal agriculture programs, such as commodity-based price and income support and rural development programs.

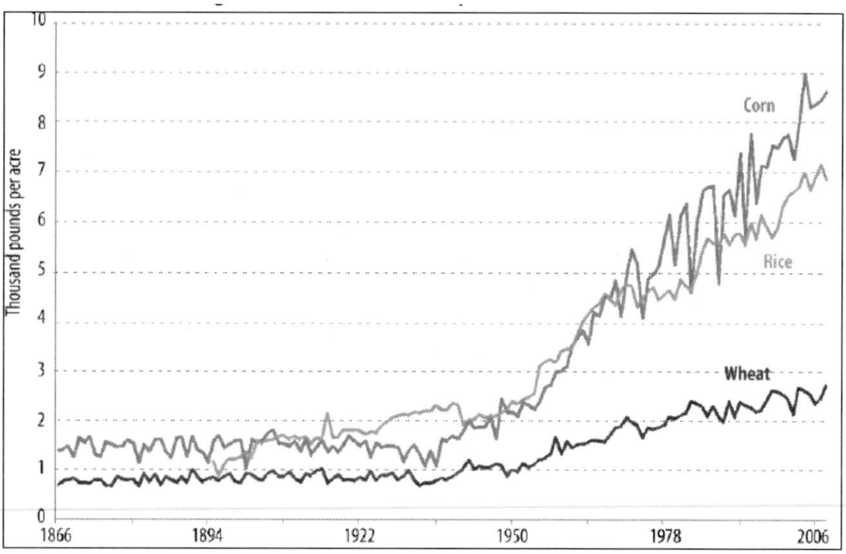

Source: Philip Pardey, "Putting U.S. Agricultural R&D and Productivity Developments in Perspective," presentation to Farm Foundation Conference, Agricultural Research and Productivity for the Future, Washington, DC, April 28, 2009, http://www.farmfoundation.org/news/articlefiles/1705-Pardey%20.pdf.

Figure 1. U.S. Commodity Yields, 1866-2008.

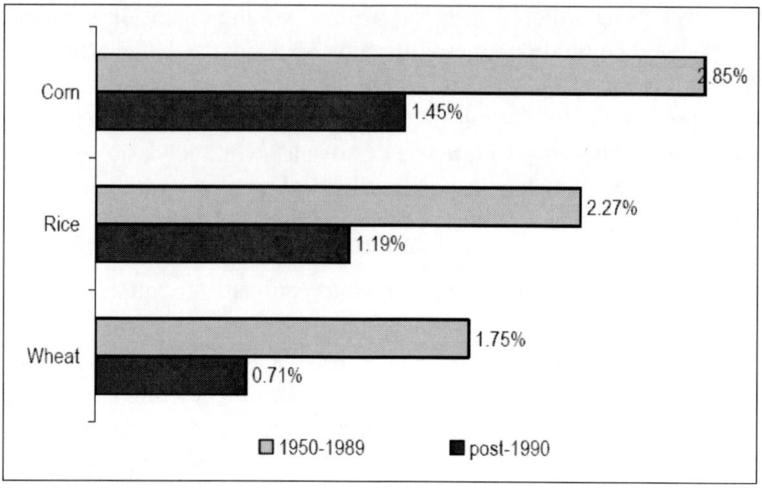

Source: Philip Pardey, "Putting U.S. Agricultural R&D and Productivity Developments in Perspective,"presentation to Farm Foundation Conference, Agricultural Research and Productivity for the Future,Washington, DC, April 28, 2009, http://www.farmfoundation.org/news/articlefiles/1705-Pardey%20.pdf.

Figure 2. Annual Yield Growth Rate, 1950-2008.

USDA Research, Education, and Economics Mission Area

The U.S. Department of Agriculture (USDA) Research, Education, and Economics (REE) mission area has the primary federal responsibility of advancing scientific knowledge for agriculture through research, extension, and education. The USDA REE agencies (see **Figure 3**) provide federal leadership in creating and disseminating knowledge spanning the biological, physical, and social sciences related to agricultural research, economic analysis, statistics, extension, and higher education. For FY2012, Congress provided $2.53 billion for the REE mission area, down $53 million from FY2011.

Four USDA agencies are responsible for carrying out the REE mission:

- **National Institute of Food and Agriculture (NIFA)**, formerly the Cooperative State Research, Education, and Extension Service (CSREES).[4] As theDepartment's primary agency that supports extramural agricultural research, education, and extension activities,

NIFA provides federal leadership inadvancing research and knowledge for agriculture, the environment, human health and well-being, and communities through grants and partnerships with the land-grant university system[5] and other organizations that work at the state andlocal level. For FY2012, NIFA received appropriations of $1.20 billion.

- **Agricultural Research Service (ARS)**. As the Department's primary intramural science agency, ARS conducts research and disseminates information related to the nation's top agricultural issues, including crop and livestock production and protection, human nutrition, food safety, rural development and natural resourcemanagement and conservation. For FY2012, ARS received appropriations of$1.09 billion.
- **Economic Research Service (ERS)**. As the Department's primary agencyresponsible for economic research and analysis, ERS conducts intramural research and provides economic and policy analysis related to food, farming,natural resource management, agricultural markets, and rural development inorder to better inform public and private decision-making. For FY2012, ERS received appropriations of $77.7 million.
- **National Agricultural Statistics Service (NASS)**. The mission of NASS, the Department's primary data-gathering agency, is to provide timely, accurate, andrelevant statistics related to U.S. agriculture. For FY2012, NASS received appropriations of $158.6 million

Although all four USDA research agencies are headquartered in Washington, DC, much of the work is executed through a set of field offices and a network of partners throughout the United States that operate at the state and local levels. Only ERS is an entirely Washington DC-based social science agency. ARS has research centers and work locations across the United States, Puerto Rico, and the Virgin Islands (**Figure 4**). NASS has offices in 45 states and Puerto Rico and a National Operations Center in St. Louis, Missouri. USDA's extramural agency, NIFA, primarily partners with state colleges of agriculture at land-grant universities in 50 states and eight U.S. territories, with their affiliated state agricultural experiment stations (SAES), schools of forestry and veterinary medicine, and the cooperative extension system.[6] The system also includes 18 historically black land-grant colleges of agriculture (the 1890 institutions) and 31 Native American colleges (referred to as tribal colleges) that gained land-grant status in 1994 (**Figure 5**).

	Appropriations $2.53 billion in FY2012	USDA states that the REE Mission Area is dedicated to the creation of a safe, sustainable, competitive U.S. food and fiber system and strong, healthy communities, families, and youth through integrated research, analysis, and education.
Intramural (federal)	**National Agricultural Statistics Service (NASS) $159 mil.**	Provides timely and accurate statistics related to U.S. agriculture (e.g., Census, crop forecasts, estimates of farm prices). DC & state offices.
	Economic Research Service (ERS) $78 mil.	Provides economic and policy analysis to inform public and private decision making related to food, farming, natural resource management, agricultural markets, and rural development. DC only.
	Agricultural Research Service (ARS) $1,095 mil. total	Conducts research and disseminates information related to crop and livestock production and protection, human nutrition, food safety, rural development, and natural resource management and conservation. Emphasis is on *national and regional problems*, including higher-risk and long-term research such as plant and animal genome program. The agency's research programs include New Products/Product Quality/Value Added; Livestock/Crop Production; Food Safety; Livestock/Crop Protection; Human Nutrition; and Environmental Stewardship. ARS has a workforce of about 6,500 full-time employees. About 100 research locations across the country, e.g., Center for Grain and Animal Health Research in Manhattan, KS.
Extramural (state with federal funding)	**National Institute of Food & Agriculture (NIFA) $1,202 mil. total** **Funding:** *Hatch Act, Evans-Allen (1890s historically black colleges), McIntire-Stennis (forestry research), Agriculture and Food Research Initiative (AFRI), Smith-Lever*	Provides federal leadership in advancing research and knowledge for agriculture, the environment, human health and well-being, and communities through grants and partnerships with the *land-grant university system* and other organizations that work at the *state and local level*. Workforce is about 350-400 employees. Formerly known as the Cooperative State Research, Education, and Extension Service (CSREES). Federal funding is provided through competitive grants and "formula funds" based on each state's farm and rural populations and matching funds from states.

NIFA Funding Destinations:

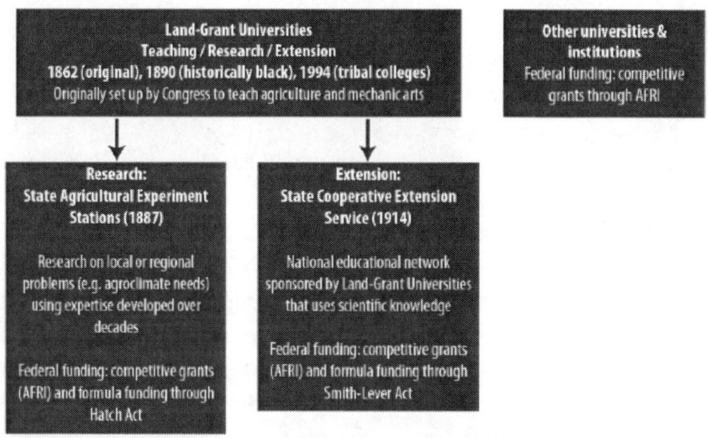

Source: CRS, using information published by USDA, http://www.usda.gov. For program details and agency descriptions, see USDA's FY2013 Budget Explanatory Notes, http://www.obpa.usda.gov/FY13explan_notes.html.

Figure 3. USDA's Research, Education, and Economics (REE) Mission Area.

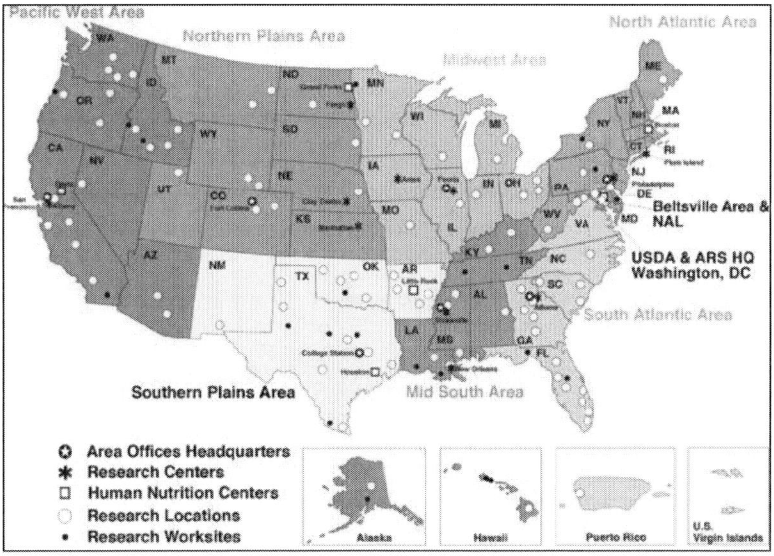

Source: USDA, Agricultural Research Service, http://www.ars.usda.gov/ pandp/ locations.htm.

Figure 4. USDA/ARS Locations.

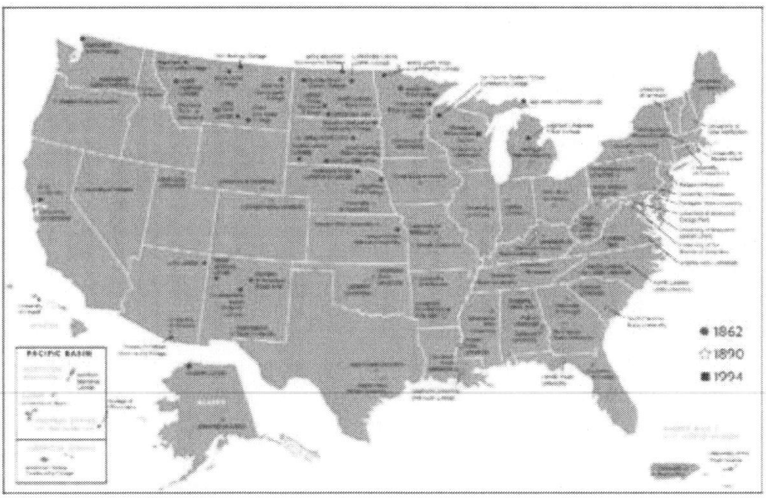

Source: U.S. Department of Agriculture, National Institute of Food and Agriculture, http://www.csrees.usda.gov/qlinks/partners/map_lgu_all_front_12_9_09.pdf.

Figure 5. Land-Grant Colleges of Agriculture.

The decentralized state-led structure for federally funded extramural activities has resulted historically in geographically specific applied research. Federally funded intramural research, on the other hand, is intended in part to address issues of national importance, and to promote basic research, regional coordination, and spillover.[7] The federal-state research system also supports USDA's regulatory programs in the areas of meat, poultry, and egg inspection, foreign pest and disease exclusion, and control and eradication of crop and livestock threats, among other things.

USDA administers extramural federal appropriations primarily through three funding mechanisms:

- **Formula Funds.**[8] Funds for research and extension are awarded to land-grant institutions (1862, 1890, and 1994 institutions), schools of forestry and schools of veterinary medicine through several formula program authorities. The amount of funds provided to each institution is determined by census-based statutory formulas. Research priorities are more geographically focused as local or regional university leaders decide which specific projects will be supported by an institution's formula grant allotment (7 U.S.C. 301 et seq.).
- **Competitive Grants**. Funds are awarded using a competitive, peer-reviewed process for fundamental and applied research, extension, and higher education activities, as well as for projects that integrate research, education and extension functions. Competitive programs are designed to enable USDA to attract a wide pool of applicants to work on agricultural issues of national interest, and to selectthe best quality proposals submitted by highly qualified individuals, institutionsor organizations (7 U.S.C. 450i(b)).
- **Noncompetitive Grant Programs**. Until recently, funds have been directed (i.e., earmarked) by Congress to support a designated institution or set of institutionsfor particular research, education, or extension topics of local, regional, or national priority. These projects have been supported through Special ResearchGrants or Direct Federal Administration Research or Education Grants (7 U.S.C. 450i(c)). Unlike during FY2008-FY2010, the enacted appropriations in FY2011and FY2012 did not include any earmarks or congressionally designated spending items for REE-related activities.

Agricultural Research Budget Overview and Trends

Congress traditionally considers research, education, and extension policy in periodic omnibus farm bills that cover virtually all USDA programs and policies. Congress focused specifically on federal agricultural research, education, and extension policies in the 1996 farm bill (P.L. 104-127) and the 2002 farm bill (P.L. 107-171), as well as in free-standing legislation enacted in 1998, (P.L. 105-185, the Agricultural Research, Extension, and Education Reform Act of 1998). The 2008 farm bill (Title VII of P.L. 110-246) reauthorized the research, education, and extension provisions of the 1996 and 1998 laws through FY2012 and contained some further revisions. The House and Senate Agriculture Committees are primarily responsible for oversight of agricultural research, education, and extension programs.

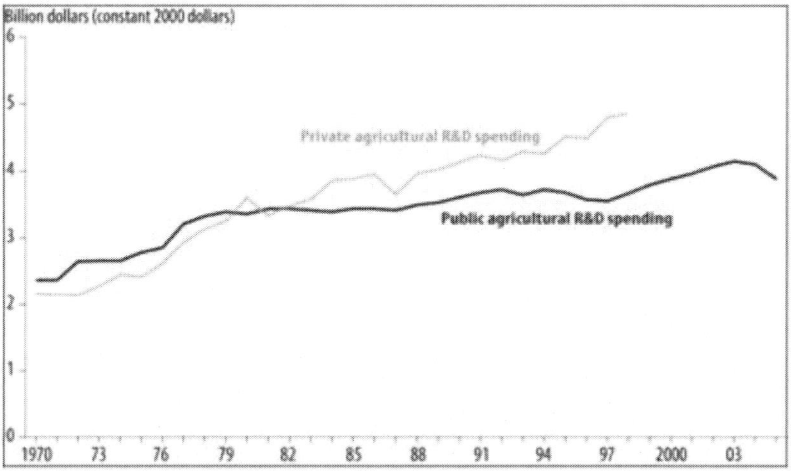

Source: David Schimmelpfennig and Paul Heisey, *U.S. Public Agricultural Research: Changes in Funding Sources and Shifts in Emphasis*, 1980-2005, Economic Research Service, USDA, Economic Information Bulletin Number 45, Washington, DC, March 2009, http://www.ers.usda.gov/Publications/ EIB45/ EIB45.pdf.

Note: Reliable estimates for total private sector investments in agricultural R&D are not available after 1998.

Figure 6. Real U.S. Public and Private Agricultural R&D Expenditures, 1970-2006.

The majority of funding for agricultural research, education, and extension activities is discretionary, and is generally provided for through annual appropriations acts. Some specific programs, such as for specialty crops, were made mandatory in the 2008 farm bill.[9]

When adjusted for inflation, public funding[10] for agricultural research grew steadily from the 1950s to the mid-1970s, remained basically level from the end of the 1970s through the 1980s, and increased only slightly during the 1990s (**Figure 6**).[11] In contrast, all public non-defense research and development spending grew considerably during that period.[12] For agriculture, there was a marked rise in public funding from 1998 through 2001, at a time of a budget surplus. One-time, supplemental funding specifically appropriated for anti-terrorism activities, not ongoing programs, was a significant factor for the increases in the FY2001-FY2003 period. Funding levels peaked in FY2010, but declined in FY2011 and FY2012 as Congress cut federal spending.

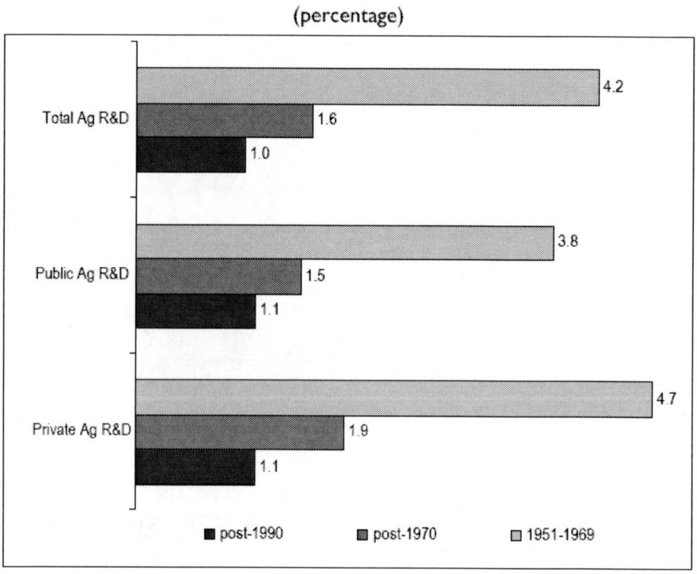

Source: Philip Pardey, "Putting U.S. Agricultural R&D and Productivity Developments in Perspective," presentation to Farm Foundation Conference, Agricultural Research and Productivity for the Future, Washington, DC, April 28, 2009, http://www.farmfoundation.org/news/articlefiles/1705-Pardey%20.pdf.

Note: Reliable estimates for total private sector investments in agricultural R&D are not available after 1998.

Figure 7. AverageAnnual Real Growth, 1950-2007.

Private-sector spending on agricultural research has continued to increase over the past several decades (**Figure 6**). The average annual growth rate (in real dollars) of agricultural spending from both the public and private sectors was highest, at over 4%, during the 1950s-1970s, and has declined considerably over time to about 1% in the most recent two decades (**Figure 7**).

The FY2012 Agriculture Appropriations Act (P.L. 112-55) contained $2.53 billion in discretionary funds for USDA agricultural research, education, and extension programs. ARS and NIFA accounted for most of the research mission area's budget, consuming about $1.09 billion (43%) and $1.20 billion (47%) of the total USDA REE budget for FY2012, respectively. This represents a 3% decrease over the FY2011 appropriation for ARS and a 1% decrease for NIFA.[13]

Over the past few decades, the growth in the research budget of USDA has lagged behind that of other national science agencies such as the National Institutes of Health (NIH), the National Science Foundation (NSF) and the National Aeronautics and Space Administration (NASA). In fact, between 1983 and 2003 the annual growth rate of USDA research spending was about one-eighth that of the NIH and about one-fourth that of all other government non-defense R&D spending overall (**Table 1**).

As a result of a relatively flat or declining USDA research budget, funding from federal agencies other than USDA has accounted for an increasing portion of total federal support for research at the state agricultural experiment stations (SAESs). Historically the USDA has been the primary federal government agency channeling funds to the SAESs. In 1970, the USDA disbursed almost 70% of the federal funds flowing to the SAESs, but by 2004 that share had declined to less than 50%, with more than half of the federal funds being disbursed by a wide range of federal agencies, including the NSF, NIH, DOE, DOD, U.S. Agency for International Development (USAID), and others. Funds from private industry for agricultural research have also been steadily increasing (**Figure 6**). Public-private partnerships can facilitate technology transfer and at the same time help to supplement federal and state support. However, some observers are concerned that both the increase in non-USDA funding and the increase in private funding might cause the focus of agricultural research to shift away from the U.S. agricultural sector's highest priorities and needs. They believe that such a shift could hamper the nation's ability to remain cutting-edge with regard to new innovations, to be competitive in a global market, and to cope with long-term challenges such as pest and disease outbreaks, climate change and natural resource management.

Table 1. Annual Real R&D Budget Growth by Federal Agency, 1983-2003

Federal Agency	Average Annual Growth Rate (percentage)
NIH	5.73
NASA	3.17
NSF	2.43
USDA	0.70
Non-Defense R&D overall	2.75

Source: USDA Research, Education, and Extension Task Force Report, 2004.

Notes: Actual spending adjusted for inflation using 2000 as base year.

OMNIBUS 2008 FARM BILL

The Food, Conservation, and Energy Act of 2008 (the 2008 farm bill, P.L. 110-246), authorized and directed the implementation of USDA's major programs through FY2012. The research title of the 2008 farm bill, Title VII, reorganized the Department's Research, Education, and Economics mission area and created a new competitive grants program for agriculture. Title VII created an umbrella coordinating entity known as the Research, Education, and Extension Office (REEO) in the Office of the Under Secretary for Research, Education, and Economics, and designated the Under Secretary as the Chief Scientist of USDA.

The 2008 farm bill also required the REEO to develop and implement a USDA Roadmap for Agricultural Research, Education, and Extension to plan and coordinate across the entire department both capacity and competitive programs, as well as USDA-administered intramural and extramural programs. In addition, all programs and authorities formerly administered by the Cooperative State Research, Education, and Extension Service (CSREES) were transferred to a newly created entity called the National Institute for Food and Agriculture (NIFA) in October 2009. The research provisions, including changes to the management and structure of REE in the 2008 farm bill, drew heavily on proposals and recommendations put forth by key stakeholder groups, including the USDA Task Force on Research, Education, and Extension and the Association of Public and Land-Grant Universities (APLU).[14]

The 2008 farm bill authorized the Agriculture and Food Research Initiative, a new competitive grants program for basic and applied research administered by NIFA. AFRI expands and replaces the National Research Initiative (NRI) Competitive Grants Program and incorporates and replaces the former Initiative for Future Agriculture and Food Systems (IFAFS), both of which the 2008 farm bill repealed. The Under Secretary is required to submit a unified annual budget covering all activities of the REEO and NIFA.

The farm bill authorized annual appropriations of $700 million for AFRI, representing the combined level of authorized and mandatory funding that the NRI and IFAFS were authorized to receive in previous years.[15] The 2008 farm bill also provided a total of $333 million in mandatory funds over five years for (1) a new specialty crop research initiative ($230 million), (2) research on fresh produce food safety ($25 million), and (3) organic agriculture research ($78 million). Annual appropriations of such sums as necessary for research, education, and extension programs were much the same as in the previous farm bill.

In addition, the research title of the enacted farm bill included major initiatives to provide capacity-building support to Hispanic-serving agricultural colleges and to make them eligible to receive funding through a wider range of grant programs.

The following subsections provide additional detail on the significant changes made to the USDA REE mission area in the enacted 2008 farm bill.[16]

The REEO and the Roadmap

Section 7511of the 2008 farm bill required the Under Secretary, designated as the Chief Scientist of the Department, to establish the Research, Education, and Extension Office (REEO) "to coordinate research programs and activities of the Department of Agriculture." Located in the Office of the Under Secretary for Research, Education, and Economics, the REEO is designed to be the central office within USDA where science priorities are set, research activities are coordinated and prioritized, and scientific capacity is strategically aligned. The REEO is organized into six divisions: (1) renewable energy, natural resources and environment; (2) food safety, nutrition and health; (3) plant health and production; (4) animal health and production; (5) agricultural systems and technology; and (6) agricultural economics and rural communities. Each division has its own division chief. The REEO is to identify key science issues and needs, coordinate and prioritize research activities and programs, and strategically align the department's scientific capacity. The division chiefs are expected to work with the National Agricultural Research, Extension, Education, and Economics Advisory Board to coordinate all of the mission area's activities across the department, including intramural research (ARS, ERS, NASS) and extramural research (NIFA).

Section 7504 of the 2008 farm bill required that the REEO, under the leadership of the Under Secretary, prepare a Roadmap for Agricultural Research, Education, and Extension to identify current trends, constraints, gaps, and major opportunities that no single entity within the USDA would be able to address individually. The reorganization of REE and the Roadmap were based largely on a proposal developed by APLU, *Create Research, Extension, and Teaching Excellence for the 21st Century* (CREATE-21), a comprehensive recommendation for reorganizing the REE system after a nationwide deliberative process within the land-grant system.

National Institute of Food and Agriculture (NIFA)

Section 7511 of the 2008 farm bill established the National Institute of Food and Agriculture (NIFA), in an attempt to streamline and reinvigorate national investment in agricultural research. In October 2009, NIFA assumed all authorities and programs from CSREES, which was responsible for funding USDA extramural research. Through the creation of NIFA, Congress intended to "increase the visibility of competitive programs at USDA's research agencies,"[17] and to create an agency equal in stature to other federal science grant-making agencies such as the National Institutes of Health (NIH) and the National Science Foundation (NSF).

Changes in the funding, structure, and priorities of USDA REE, and the vision for a NIFA, drew heavily from recommendations made by several stakeholder groups, including APLU and USDA itself. The 2002 farm bill authorized USDA to create a task force to prepare a report examining the merits of establishing national institutes in one or more of the agricultural sciences.[18] The creation of a new separate grant-making agency within USDA that was solely responsible for administering competitive grants programs in agricultural research and extension was one of the recommendations that came out of a National Academy of Sciences 2000 report looking at the efficacy of the National Research Initiative Competitive Grants Program.[19] In July 2004 the USDA task force published its report, which advocated for the concept of a new NIFA.[20] The task force modeled NIFA on NIH and NSF, recommending that NIFA should accomplish its mission primarily through administering competitive peer-reviewed grants that support and promote high-caliber, fundamental agricultural research. It recommended that the head of NIFA should be a distinguished scientist, appointed by the president for a six-year term, with the potential for one reappointment. In addition, the task force recommended that NIFA distribute research grants through a competitive, peer-reviewed process with a budget that would increase over a five-year period until it reaches $1 billion per year.

Enhanced Competitive Grant Funding: Agriculture and Food Research Initiative (AFRI)

Section 7406 of the 2008 farm bill amended the Competitive, Special, and Facilities Research Grant Act to establish the Agriculture and Food Research Initiative (AFRI), which is designed to be the flagship competitive grants program at USDA and is housed at NIFA. AFRI replaces two other USDA grant programs, the Initiative for Future Agriculture and Food Systems (IFAFS), which emphasized more applied research, and the National Research Initiative (NRI) competitive grants program, which emphasized fundamental, or basic research. Both of these programs were eliminated in the 2008 farm bill. AFRI provides competitive grants for basic and applied research, education and extension to colleges and universities, agricultural experiment stations, and other organizations conducting research in priority areas of food and agriculture including plant health and production; animal health and production; food safety, nutrition, and health; renewable energy, natural resources, and environment; agriculture systems and technology; and agriculture economics and rural communities.

AFRI was authorized to receive appropriations of $700 million annually in the 2008 farm bill. However, actual appropriations were $201.5 million in FY2009, $262.5 million in FY2010, and $264.5 million in both FY2011 and FY2012. The 2008 farm bill mandated that AFRI allocate 60% of grant funds for basic research and 40% for applied research. At least 30% of total funds must be used to integrate research with education and/or extension activities.

Other Farm Bill Research Initiatives

Specialty and Organic Crops Research

Funding for specialty (fruits and vegetables) and organic crops research was extended and expanded in the 2008 farm bill.[21] Supporters of specialty crops, including producers, processors, and trade associations, argued in favor of increasing federal expenditures for specialty crops research. They argued that growth in the value of production of specialty crops has not been matched by commensurate growth in public agricultural research spending. In addition to expanding discretionary funded programs, Congress also increased the use of mandatory funds specifically for specialty crop and organic research.

- **Specialty Crops Research Initiative**. The farm bill included $230 million in mandatory funds ($30 million for FY2008, and $50 million annually thereafter for FY2009-FY2012) for a specialty crops research initiative over five years and authorized annual appropriations of $100 million. Five research areas are prioritized, including crop improvement; identification of pest and disease threats, production efficiency; improved technologies; and prevention and detection of food safety hazards. Each area is to receive at least 10% of the grant funds. Grant recipients must provide a 100% match either in funds or in-kind support. The program gives priority to projects that are multistate, multi-institutional, or trans-disciplinary, and includes explicit mechanisms to communicate results to producers and the public.
- **Organic Agriculture Research and Extension Initiative**. Originally authorized under the 2002 farm bill at $3 million per year in mandatory funding, the Organic Agriculture Research and Extension Initiative funds research and extension programs that are intended to enhance the ability of producers and processorswho have already adopted organic standards to grow and market high qualityorganic agricultural products. This program emphasizes research and outreach to assist farmers and ranchers with farm planning and ecosystem integration, and aims to deliver applied production information to producers. The 2008 farm billincreased the mandatory funding level from $3 million annually to $18 million for FY2009, and $20 million annually thereafter in FY2010-FY2012. The 2008 farm bill also included an authorization for an additional $25 million in discretionary funds each year.

Bioenergy Research

Three agricultural energy provisions that extend and expand existing programs, as well as create a new competitive grants program for on-farm biomass crop research, were included in the research title of the 2008 farm bill.[22] These programs provide grants to develop innovative methods and technologies for the economic and efficient conversion of agricultural waste to bioenergy, support the dissemination and implementation of biobased energy technologies, and encourage collaboration between the USDA, DOE, and land-grant universities.

- **Nutrient Management Research and Extension Initiative**. This program,originally authorized in the 1990 farm bill (P.L. 101-624), provides matchinggrants for finding innovative methods and technologies for economic use and/ordisposal of agricultural waste. The 2008 farm bill amended this program to include the production of renewable energy from animal waste as an eligibleactivity, and extended the program through 2012, authorizing such sums as necessary annually.
- **Agricultural Bioenergy Feedstock and Energy Efficiency Research and Extension Initiative**. This new program awards competitive matching grants (up to 50%) for projects with a focus on supporting on-farm biomass crop research and the dissemination of results to enhance the production of biomass energycrops. Discretionary appropriations of $50 million annually were authorized for FY2008-FY2012.
- **Sun Grant Program**. This program was added to the Energy Title (Title IX) subsequent to the 2002 farm bill under the Sun Grant Research Initiative Act of2003, and established five national Sun Grant research centers at land-grant universities, each covering a different national region, to enhance coordination and collaboration between USDA, DOE, and land-grant universities in the development, distribution, and implementation of biobased energy technologies. The 2008 farm bill moved this provision to the research title, established a sixth regional center—Western Insular Pacific Sub-Center at the University of Hawaii—and authorized $75 million per year for FY2008-FY2012.

Pollinator Protection Research

The 2008 farm bill amended the Food, Agriculture, Conservation and Trade Act of 1990 by authorizing three new pollinator protection programs for FY2008-FY2012, and repealing a previously enacted honey bee disease research program.[23] The 2008 farm bill authorized $10 million a year for research and extension grants that address pollinator biology, ecology, immunology, and genomics, including factors affecting colony collapse disorder in honeybees. An additional $7.25 million annually was included to increase the capacity and infrastructure of USDA to address pollinator health issues and to conduct in-house research on colony collapse order at USDA facilities. The 2008 farm bill also authorized $2.75 million annually for a national honey bee pest and pathogen surveillance program.

CURRENT ISSUES

New Funding for Agricultural Research

A recurring policy issue is whether or not there is a need for more federal spending for agricultural research, education, and extension. The current fiscal and political situation will likely result in debate and discussion, especially if Congress considers the creation of new programs, expansion and/or scaling back of existing programs, changes in mandatory program spending,[24] and congressional earmarks. Some argue that the stagnant growth in inflation-adjusted USDA funding for agricultural research, education, and extension over the past few decades has hurt the ability of the U.S. agricultural sector to stay productive and competitive. It is widely acknowledged that new innovations and technologies related to production, processing, marketing, and natural resource management are essential for continued productivity gains and economic growth of the sector.

Critics argue that USDA has not been successful at elevating agricultural research to the same priority level with policymakers as other sectors such as health, and that U.S. agriculture will suffer over the long term because of a lack of new innovations. These critics argue that the lack of public investment in new agricultural innovations will have dire consequences in the future, especially given new and varied challenges, such as rising production costs, especially for fuel and inputs, new pest and disease outbreaks, increasing frequency of extreme weather events such as drought and floods, and climate change.

On the other hand, some argue that the federal government should have a limited role in funding agricultural research, and that taxpayer dollars should not be used to support what should be a private sector endeavor. In addition, due to a severely constrained federal budget in recent years, there are limited resources available with which to support the agricultural sector. Historically, Congress has not prioritized increasing funding for agricultural research, education, and extension activities, and instead has tended to fund programs designed to provide more immediate benefits to farmers, such as income support and safety net programs. Some believe that farmers would have to wait too long to see the payoff from research activities, and that it is difficult to justify the investment when farmers typically need federal support to address their needs in the short term. Others believe that the states and the private sector should fill the research funding gap left by the federal government.

While some in Congress have attempted to provide alternative sources of funding for research programs, such as authorizing $600 million in mandatory funding in the Agriculture Research, Extension, and Education Reform Act of 1998 for a competitive research grants program, Congress has blocked mandatory funds from being spent on new competitive agricultural research grants every year since 2002.

At the same time, while private sector funding has increased over time to fill some of the gap in public spending, there is growing concern among some that private sector funding focuses primarily on taking existing technologies to market (i.e., more applied research), and does not focus on basic problems and/or longer-term challenges that the agricultural sector may face in the future such as environmental sustainability or adaptation to climate variability.

Funding Mechanism: Formula Funds vs. Competitive Grants

Much debate has surrounded the implications of various funding mechanisms for agricultural research, in particular federal formula funding vs. external peer-reviewed competitive grant funding. Efforts to improve public agricultural research efficiency have often included calls to increase the use of competitive grants as a means of more effectively allocating limited federal resources. Two historically influential reports published by the National Academy of Sciences (NAS)[25] and the Rockefeller Foundation[26] argued that the agricultural research of 30-40 years ago had become overly focused on applied research rather than cutting edge basic research, and both reports recommended a shift in funding mechanisms toward more use of competitive funding rather than formula funding of the state agricultural experiment stations (SAESs).

The USDA differs from other federal research agencies in allocating the majority of its annual research appropriation directly to in-house research agencies (ARS, ERS, and NASS), and in allocating the majority of its extramural funding to state agricultural colleges and land-grant universities through statutory formula funds. In contrast to USDA's research and extension funding mechanisms, the federal government's two largest science agencies, the National Science Foundation (NSF) and the National Institutes of Health (NIH) allocate the majority of their appropriations through competitively awarded grants to scientists in a variety of research venues nationwide.[27] In another report published in 1989, the NAS recommended that at least 35% of total USDA research funding be distributed competitively, and called for the expansion of USDA's competitive grants programs.[28] The 1989 NAS report also recommended that expanded support for competitive programs should come from new money and specifically advised against redirecting funds from ARS or the Hatch and Smith-Lever Acts, which are the primary and longest-standing formula fund provisions for cooperative research and extension, respectively. When the NAS made its recommendation in 1989, it determined that less than 6% of USDA's research funding was competitively awarded. In FY2008, it was approximately 14%, according to CRS calculations.[29]

Formula funds represent the largest allocation of USDA extramural funds, but the relative and absolute amount of resources distributed by formula funds has been steadily declining over the past three decades (**Figure 8**). The decrease in formula funds has been met with a commensurate increase in allocation to competitive grants and special grants.

Previous Congressional Action on Competitive Agricultural Research Funding

Following publication of the 1989 NAS report, Congress expanded USDA's existing competitive research grant program in the 1990 farm bill (P.L. 101-624). Authority for competitive grants originally had been established in the 1977 farm bill (P.L. 95-113), but the program remained small in comparison to the direct and formula-funded research programs. Congress authorized annual appropriations for the expanded program, the National Research Initiative (NRI), of $150 million in FY1991, $275 million in FY1992, $350 million in FY1993, $400 million in FY1994, and $500 million in FY1995. Congress subsequently reauthorized the NRI in the 1996 farm bill, the 1998 agricultural research law, and the 2002 farm bill, each time renewing a $500 million authority for annual appropriations. Yet, NRI funding rarely exceeded the appropriated level of $182 million in FY2008, the last year of the program's existence.

In 2006, President George W. Bush proposed eliminating federal formula funding of agricultural research and replacing it with increased funding for NRI, and a new competitive grants program for the SAES system. Together, the NRI increase and the new grant program would have constituted a 14% increase in federal funding for state research over the previous year. At the state level, however, the proposal would have represented a 50% cut in funding for core programs. APLU opposed the President's proposal and Congress did not include it in the final FY2006 appropriations act (P.L. 109-97). In FY2007 the Bush Administration made another attempt to alter the allocation of formula funds in favor of a more competitive process, but these proposals were again rejected by Congress.

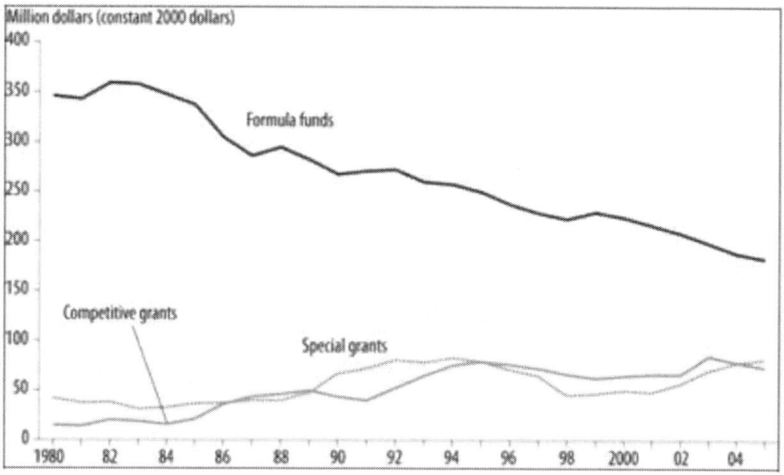

Source: David Schimmelpfennig and Paul Heisey, *U.S. Public Agricultural Research: Changes in Funding Sources and Shifts in Emphasis*, 1980-2005, Economic Research Service, USDA, Economic Information Bulletin Number 45, Washington, DC, March 2009, http://www.ers.usda.gov/Publications/ EIB45/ EIB45.pdf.

Notes: Actual spending adjusted for inflation using 2000 as base year.

Figure 8. USDA Extramurally Administered Funds,1980-2005.

With the 2008 farm bill, Congress strengthened the competitive process for agricultural research. The establishment of the National Institute of Food and Agriculture (NIFA) as the primary agricultural research agency within the USDA and the new flagship competitive grants program that NIFA administers, called the Agriculture and Food Research Initiative (AFRI), are the outcome of ongoing efforts to elevate the role of competitive grants as an important federal funding mechanism for agricultural innovation.

Implications of Funding Mechanism

The choice of funding mechanisms is viewed by some as important because it is thought to determine where and by whom the research is conducted, and the type of research performed. On the one hand, the competitive, peer-reviewed process is thought to have an advantage because a wider pool of candidates is eligible to apply for funding (e.g., grant recipients are not limited to land-grant institutions or SAESs), and it is thought to engage the best and brightest minds in addressing challenges facing the agriculture sector. At the same time, it is widely acknowledged in the agricultural community that USDA-funded research, whether carried out intramurally or through formula funds, also has an important role to play. Using a census-based formula to calculate the annual distribution of research funds to each state has meant that the state allocations have been quite constant from year to year, since annual appropriations have remained level or increased slightly. Although all federal sources account for 30% or less of total funding for the experiment stations (including grants from non-research agencies within USDA and from other federal departments), the reliability of the formula funds has resulted in them traditionally being used to support the core ongoing research programs of the state agricultural experiment stations, which underpin academic programs at many universities.

Studies have shown that funding through competitive grants tends to favor basic research, reach a greater proportion of non-land grant universities, and are concentrated among fewer states than funding that is allocated by statutory formula funds. States with large agricultural production and top-ranked academic programs in biology and agricultural sciences were generally more competitive and more successful in receiving larger shares of federal funds allocated as competitive grants.[30]

Other studies have shown that federal formula funding has a larger impact on agricultural productivity over the longer term than federal competitive grants and contracts.[31] The rationale is that federal-level research is steady funding that can support core or foundation research and is best able to take on higher-risk and long-term projects of national importance, such as deciphering plant and animal genomes, conducting longitudinal studies on human nutrition, and measuring and analyzing current and historical socioeconomic factors in the U.S. food and fiber sector. Proposals that address problems of concern to an entire state or region, and/or are multi-disciplinary, are typically underfunded in a national competitive-grant process, despite the fact that such research problems are considered by many to be of critical concern and may have a large net social payoff to the agricultural sector.

Intramural vs. Extramural Funding

ARS is the principal in-house or intramurally funded research arm of the USDA. Many believe that maintaining some level of federally funded internal research allows ARS to fill an important niche, not met by industry or other institutions, specifically to address research problems of national and long-term priority such as conservation and improvement of plant genetic resources, surveillance and monitoring of national and regional disease outbreaks, soil and water resource management, and adaptation to increasing climate variability and extreme events. At the same time, while intramural funding is intended to cover all costs needed to conduct mission research, funding levels for intramural research have been relatively flat over the past decade and some feel that funding levels are not sufficient. ARS is allowed to accept funds from outside sources, including other USDA agencies; other federal agencies; state, county, and municipal agencies and institutions; and colleges and universities. USDA leadership has recently been encouraging ARS scientists to leverage extramural funding and partnerships where appropriate.

Some believe that in order for ARS to maintain strong, relevant research programs, ARS scientists should pursue extramural partnerships and sources of funding to carry out their work. Others express concern that pursuing extramural funding may be a distracting and time-consuming endeavor for ARS scientists, which would take away focus from ARS's primary national program priorities. There is also concern over setting specific quantitative goals or unrealistic expectations for extramural funding of ARS research. In addition, some believe that ARS scientists have an unfair advantage in competing with other agricultural scientists, who do not have an endowed source of support for core research expenditures.

Earmarks

One of the most controversial sources of funding for ARS, NIFA, and the land-grant agriculture schools are earmarked grants in annual agricultural appropriations acts. For FY2009, Congress appropriated over $159 million for 102 ARS research projects at various labs across the nation, and over $122 million for 254 designated research and extension projects at various land-grant institutions administered by NIFA. In FY2010, earmarks totaled $115 million for ARS and $135 million for NIFA. The FY2011 and FY2012 enacted appropriations did not include any earmarks or congressionally designated spending items for REE-related activities.

Congressionally designated grants typically do not receive any type of formal review before they are awarded. However, in response to repeated criticism that such funds could be supporting weak or unnecessary research, NIFA (formerly CSREES) developed many years ago an internal merit review process that research projects must undergo before the funds will be released. Merit reviews may result in improvements being made to the project (in research methodology, for example) before the funding is transferred to the designated institution.[32]

Under appropriations rules for the 112th Congress, Congress banned earmarks. Prior to that, controversy over earmarked appropriations had arisen nearly every year. Each annual budget request had proposed eliminating the designated grants under both ARS and NIFA, stating that they deflect money away from higher-priority research in the national interest. Supporters of such grants countered that Members are most qualified to know what research is most important to their districts and to make sure those local needs are not ignored in the larger research picture.

The Federal Budget and Agricultural Research in the Next Farm Bill

The federal budget continues to be severely constrained, which makes finding new money for agricultural research a difficult endeavor. While the 2008 farm bill provided a significant funding boost for agricultural research, relatively speaking, it is unclear whether budgetary resources and political will can sustain funding for agricultural research and related activities.

Several mandatory programs that were authorized in the 2008 farm bill do not have a budget baseline that extends beyond the end of the 2008 farm bill

(September 30, 2012).[33] If policymakers want to continue these programs in the next farm bill, they will need to pay for them with other offsets. Of the 37 mandatory farm bill programs that have no budget baseline after the 2008 farm bill expires, three were authorized in the Research title: the Specialty Crop Research Initiative ($230 million over five years); the Organic Agriculture Research and Extension Initiative ($78 million over five years); and the Beginning Farmer and Rancher Development Program ($75 million over five years). In addition, eight mandatory programs in the Energy Title (Title IX), totaling about $1.9 billion over five years, also do not have budget baseline going into the next farm bill. Some of these programs, such as the Biomass Research and Development Program, have research objectives and are administered by NIFA.[34]

Finally, there has been interest in Congress and the Administration in increasing resources for an international agricultural development agenda to reduce global hunger and poverty, particularly in sub-Saharan Africa. Some key areas that have been identified include U.S. support for international agricultural research and education, and U.S. support for technical capacitystrengthening abroad.

Some believe that the convergence of several factors, including the constrained federal budget, ongoing international trade negotiations, the growing importance of non-program, specialty crops to the U.S. agricultural economy, and the potential role of international agricultural development as a key pillar of U.S. foreign affairs strategy, could foster congressional consideration of increasing federal support for U.S. agriculture through agricultural research, education, and extension funding beyond the farm bill.

End Notes

[1] Keith O. Fuglie and Paul W. Heisey, *Economic Returns to Public Agricultural Research*, USDA Economic Research Service, Economic Brief Number 10, Washington, DC, September 2007, http://www.ers.usda.gov/Publications/EB10/.

[2] J. M. Alston, C. Chan-Kang, and M. C. Marra et al., *A Meta-Analysis of Rates of Return to Agricultural R&D*, International Food Policy Research Institute, Research Report 113, Washington, DC, 2000, http://www.ifpri.org/sites/ default/files/publications/rr113.pdf. This analysis showed that the actual rate of return is highly dependent upon many factors including the specific commodity, the type of research conducted (basic vs. applied), and the duration of the research investment among other things.

[3] Philip Pardey, "Putting U.S. Agricultural R&D and Productivity Developments in Perspective," presentation to Farm Foundation Conference, Agricultural Research and Productivity for the Future, Washington, DC, April 28, 2009, http://www.farmfoundation.org/ news/ articlefiles/1705-Pardey%20.pdf.

[4] The Food, Conservation, and Energy Act of 2008 (P.L. 110-246, the 2008 farm bill) established the National Institute of Food and Agriculture (NIFA), which is under the jurisdiction of the REE mission area. All authorities previously administered by CSREES were transferred to NIFA (Section 7511 of P.L. 110-246). NIFA is covered in more detail in the "Recent Legislative Action" section of this report.

[5] The term "land-grant" refers to the law first establishing an institution of public higher education in each state to teach the "agricultural and mechanical arts." The Morrill Act of 1862 gave a grant of federal land to each state and directed the state to sell the land and use the proceeds to establish a college of agriculture. In many states, the original 1862 school became the foundation for the state university, growing to include a wide range of academic disciplines, including agriculture. These large institutions sometimes are referred to as "land-grant universities," but USDA funding and programs pertain only to the colleges of agriculture within them.

[6] The cooperative extension system is a nationwide, non-credit educational network where each U.S. state and territory has a state office at its land-grant university and a network of local or regional offices.

[7] Economists use the term "spillover" to capture the idea that some of the economic benefits of research and development activities affect agents and locales beyond where the research was performed.

[8] Federal "formula funds" provide a major source of funding for public research and extension activities at land-grant institutions, including the 1862, 1890, and 1994 land-grant institutions, and are appropriated to the states on the basis of statutory formulas that change infrequently. Formula funds for research at the state agricultural experiment stations (SAESs) are authorized under the Hatch Act of 1887, which Congress amended in 1955 to include a formula that distributes the federal appropriation among states based on each state's farm and rural population as determined by the U.S. Census (7 U.S.C. 301). The Hatch Act, as amended, also requires dollar-for-dollar matching funds from state appropriations; most states, however, appropriate three to four times the federal allotment. The Hatch Act also requires each state to use 25% of its Hatch Act funds to support multi-state or regional research. The Smith-Lever Act provides federal formula funds through NIFA for cooperative extension activities using statutory formulas and non-federal matching requirements similar to the Hatch Act (7 U.S.C. 341). Federal funding supporting forestry and veterinary programs at the land grant institutions also is distributed among the institutions according to formulas, but these have different criteria than the Hatch Act and Smith-Lever Act formulas. In addition, McIntire-Stennis formula funds support state designated institutions' cooperative forestry research programs (16 U.S.C. 582a et seq.). Animal Health formula funds support research into the prevention and control of animal diseases that affect agricultural productivity (7 U.S.C. 3195).

[9] For more discussion of mandatory funding for agricultural research and the 2008 farm bill, see the "Other Farm Bill Research Initiatives" section of this report.

[10] Public funding in this figure includes research funded by USDA intramurally (ARS) and extramurally (NIFA,formerly CSREES), other non-USDA federal agencies (i.e., NIH, DOE, NSF etc), and state appropriations.

[11] David Schimmelpfennig and Paul Heisey, *U.S. Public Agricultural Research: Changes in Funding Sources and Shifts in Emphasis*, 1980-2005, Economic Research Service, USDA, Economic Information Bulletin Number 45,Washington, DC, March 2009, http://www.ers.usda.gov/Publications/EIB45/EIB45.pdf.

[12] American Association for the Advancement of Science (AAAS) R&D Budget and Policy Program,http://www.aaas.org/spp/rd/histda13.pdf.

[13] For more detailed analysis of Agriculture appropriations, see CRS Report R41964, *Agriculture and Related Agencies: FY2012 Appropriations*.

[14] APLU was previously known as National Association of State Universities and Land Grant Colleges (NASULGC); NASULGC changed its name to APLU in March 2009.

[15] AFRI annual authorization of $700 million discretionary spending essentially combines NRI discretionary funding (authorized at $500 million annually) with IFAFS mandatory funding ($200 million annually). NRI was appropriated $182 million in FY2008. IFAFS was established by the Agricultural Research, Extension, and Education Reform Act of 1998 (P.L. 105-185) as a five-year, $600 million, competitive grants program supported by mandatory funds made available from the USDA's Commodity Credit Corporation, instead of appropriated funds. Congress reauthorized the initiative in the 2002 farm bill, raising its authorized funding level from the original $120 million annually to $140 million in FY2004, $160 million in FY2005, and $200 million annually thereafter. USDA awarded grants under the initiative in FY2000 and FY2001. In FY1999 and every year since FY2002, Congress has blocked CCC funds from being spent on the initiative except to service existing grants until they end. Both NRI and IFAFS were eliminated in the 2008 farm bill.

[16] For a more complete examination of the issues and legislative proposals, see CRS Report RL34352, *Agricultural Research, Education, and Extension: Farm Bill Issues.*

[17] S.Rept. 110-220.

[18] The FY2003 Emergency Supplemental Appropriation Act (P.L. 108-11) authorized $499,000 from the ARS budget to support the study.

[19] National Academy of Sciences, *National Research Initiative: A Vital Competitive Grants Program in Food, Fiber, and Natural-Resources Research*, Washington, DC, 2000, http://www.nap.edu/catalog/9844.html.

[20] *National Institute for Food and Agriculture: A Proposal, report of the Research, Education, and Economics Task Force of USDA*, July 2004. The report is available at http://www.ars.usda.gov/sp2userfiles/place/00000000/ national.doc.

[21] For a more complete examination of specialty crops issues and legislative proposals, see CRS Report RL33520, *Specialty Crops: 2008 Farm Bill Issues.*

[22] For a more complete examination of bioenergy issues and legislative proposals, see CRS Report RL34239, *Biofuels Provisions in the 2007 Energy Bill and the 2008 Farm Bill: A Side-by-Side Comparison.*

[23] For more about pollinator issues including colony collapse disorder, see CRS Report RL33938, *Honey Bee ColonyCollapse Disorder.*

[24] CRS Report R41245, *Reductions in Mandatory Agriculture Program Spending.*

[25] National Academy of Sciences (NAS), *Report of the Committee on Research Advisory to the U.S. Department ofAgriculture*, National Academy Press, Washington, DC, 1972.

[26] Rockefeller Foundation, *Science for Agriculture*, The Rockefeller Foundation, New York, NY, 1982.

[27] It should be noted that although the majority of NIH funding is competitively awarded, the NIH also has a large in-house research component.

[28] National Academy of Sciences, *Investing in Research: A Proposal to Strengthen the Agricultural, Food, and Environmental System* (Washington, DC: National Academy Press, 1989).

[29] Computed to include funding for the NRI competitive grants program, competitively awarded extension grants, integrated research and extension program, and outreach programs for socially disadvantaged farmers, divided by total funding for ARS and CSREES for FY2008.

[30] Kelly Day Rubenstein, Paul W Heisey, and Cassandra Klotz-Ingram et al., "Competitive Grants and the Funding of Agricultural Research in the United States," *Review of Agricultural Economics*, vol. 25, no. 2 (September 24, 2003), pp. 352-368.

[31] Wallace E. Huffman and Robert E. Evenson, "Do Formula or Competitive Grant Funds Have Greater Impact on State Agricultural Productivity," *American Journal of Economics*, vol. 88, no. 4 (November 2006), pp. 783-798.

[32] See policies and procedures for ARS Reimbursable and Trust Fund Agreements, at http://www.afm.ars.usda.gov/ppweb/pdf/324-0.pdf.

[33] For more information, see CRS Report R41433, *Expiring Farm Bill Programs Without a Budget Baseline*.

[34] For a discussion of issues across all titles of the farm bill, see CRS Report R42357, *Previewing the Next Farm Bill*.

In: U.S. Agricultural Preparedness …
Editor: Warren C. Young

ISBN: 978-1-62618-437-4

Chapter 3

THE COMPLEMENTARY ROLES OF THE PUBLIC AND PRIVATE SECTORS IN U.S. AGRICULTURAL RESEARCH AND DEVELOPMENT*

U.S. Department of Agriculture

INTRODUCTION

This brief examines the funding and performance of agricultural research and development (R&D) to assess the evolving roles of the public and private sectors in the U.S. agricultural research system. There is a clear long-term trend toward greater private-sector funding and performance of R&D. In 2007, the private sector performed 53 percent of total food and agricultural research in the United States, and privately funded R&D has grown faster than publicly funded R&D over the long term (figure 1).

- The increase in agricultural R&D conducted by the private sector reflects the increasing prevalence of public-private partnerships and other policies to enhance private returns on R&D, and the private

* This is an edited, reformatted and augmented version of a United States Department of Agriculture Economic Research Service publication, Economic Brief Number 19 dated September 2012.

companies' incentive to capitalize on new opportunities for innovation from prior public investments in the agricultural sciences.

- Most of the growing share of private-sector agricultural R&D supports researchers working in private companies who focus on topics and issues with the highest expected private returns. Relative to the public sector, this leads to a smaller portfolio of research topics and a greater emphasis on short-term research.
- Public-sector funders and performers of R&D play a largely complementary role by emphasizing social returns in the selection of research topics and valuing rapid and widespread disclosure of new knowledge.

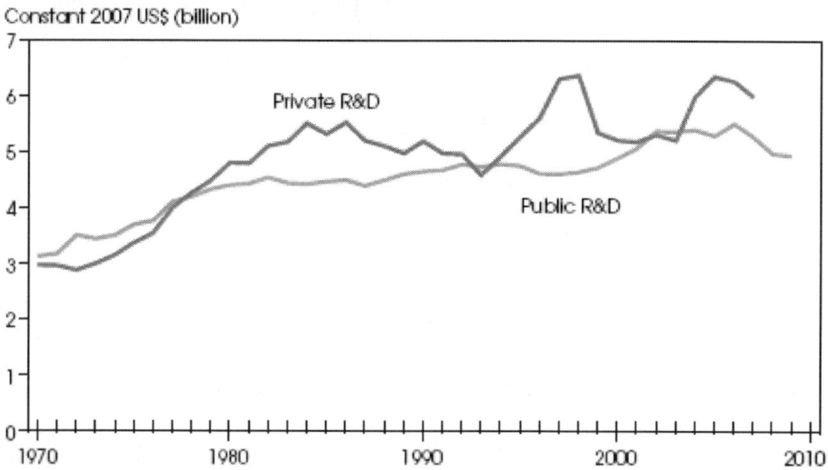

Source: Fuglie et al., 2011.

Figure 1. Long-term growth in funding for U.S. food and agriculture research and development (R&D) is higher in the private sector.

PUBLIC AND PRIVATE FUNDS SUPPORT RESEARCH AT DIFFERENT INSTITUTIONS

The U.S. agricultural research system receives funding from three primary sources: (1) Federal appropriations, (2) State appropriations, and (3) nongovernment sources (chiefly private-sector firms but also charitable foundations). These funds support research that is carried out by three main types of institutions: (1) Federal research facilities; (2) State-affiliated land-grant universities, State agricultural experiment stations (SAES), and other cooperating institutions such as schools of forestry, veterinary schools, and 1890 and 1994 land-grant institutions; and (3) private laboratories and experimental sites, nearly all of which are owned and operated by for-profit firms (fig. 2). Private institutions include firms in the food manufacturing, chemical, equipment, pharmaceutical, seed, and biotechnology industries that provide inputs to the farm sector and process agricultural commodities.

Federal research facilities receive funding almost exclusively through Federal appropriations, with a small contribution (less than 1 percent) from firms through Cooperative Research and Development Agreements and other arrangements. Firms performing agricultural R&D fund nearly all of it themselves, receiving very little funding from public sources except for Small Business Innovation Research grants and research contracts. By contrast, universities, SAES, and other State institutions draw on Federal, State, and private funds to support their R&D. According to what State institutions report to USDA (2007), Federal funds from USDA and other Federal agencies typically support 50 to 60 percent of the agricultural R&D at State institutions, and State-level appropriations support roughly another third.. Non-government sources of funds—including research contracts from private firms, technology licensing fees, and foundation grants—provide the remaining 20 percent of research funding for these State institutions.

Public and Private R&D Differ in Goals, Specialization

Private firms invest in R&D to earn the highest possible expected *private returns*, the economic benefits from new products and technologies that firms can appropriate as profit. But private returns do not fully reflect the benefits from R&D to customers, consumers, or competitors. *Social returns* to R&D investment include both the firm's private returns as well as economic benefits to these other groups in society.

The argument that private returns to R&D do not fully reflect the benefits of technological progress to society suggests the private sector underinvests in R&D from a societal perspective and supports the use of public policies to encourage additional private research. Some policies provide incentives that increase private returns to R&D such as patent protection, other intellectual property rights, technical assistance, and R&D tax credits. Public funding and performance of R&D can also address private underinvestment by expanding the knowledge base to create new opportunities for innovation.

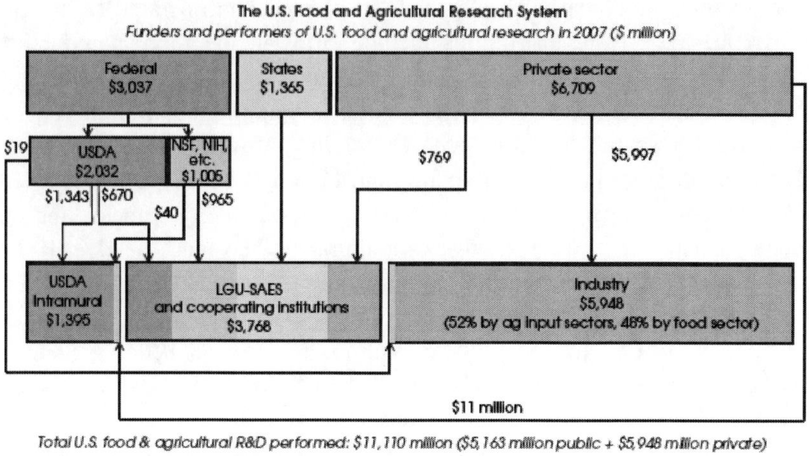

Source: Public research and development (R&D) is from U.S. Department of Agriculture (2007) except for USDA grants to industry, which is from the NSF (2009). Private R&D is from Fuglie et al. (2011). Numbers may not sum due to rounding. NSF = National Science Foundation; NIH = National Institutes of Health; LGU = land-grant universities; SAES = State agricultural experiment stations.

Figure 2. Federal, State, and private institutions both funded and performed food and agricultural research in 2007.

Social returns to R&D include an important element not reflected in private returns: the cumulative contributions of discoveries and inventions to subsequent R&D performed by other researchers in society. The potential for cumulative contributions depends on the disclosure and diffusion of new knowledge through various channels such as scientific publications. Professional norms at publicly supported research institutions encourage open, rapid publication of science, which enables other researchers to replicate, validate, and contribute to new knowledge. Universities also combine their research mission with the education and training of future scientists, which further enhances society's capacity for cumulative contributions. A study of 200 top R&D performing firms found that university research is an important source of knowledge for industrial scientific publications (Adams and Clemmons, 2008). Scientists affiliated with 110 top universities communicated research findings in nearly 200,000 articles in agriculture journals from 1981 to 1999, compared with fewer than 6,000 by scientists affiliated with 200 top R&D firms. While the market-oriented nature of private firm research may limit publication opportunities in scientific journals, private firms also have incentives to limit the availability of new technology and knowledge through trade secrets, patent protection, and other efforts to appropriate a greater share of R&D benefits. Such intellectual property protection encourages investment but can hamper social benefits to R&D from scientific exchange.

Public and Private R&D Have Different Complementary Roles

Beyond differences in how new knowledge is disclosed, public and private funders tend to invest in different "portfolios" of agricultural research topics. Private firms specialize in fewer topic areas and focus their R&D in areas with established product markets, such as food, agricultural inputs like crop seed and chemicals, farm machinery, and veterinary pharmaceuticals (fig. 3). Public institutions invest in a broader portfolio of topics than private firms. For instance, public R&D investment addresses areas such as the environment and natural resources, human nutrition and food safety, and social and community development. Research in these areas provides social benefits that are widely diffused and difficult to capture through private innovation.

Crop research and development is one area that receives a significant amount of both public and private investment, so public and private sector roles are more difficult to delineate. Public investment may crowd out private investment by supporting research that otherwise would have been done by private organizations.

Or public investment may complement private R&D by expanding the knowledge base needed to create possibilities for innovation.

The evidence favors a complementary relationship. For example, a 2001 survey of U.S. seed companies found that three-fourths of private crop breeding investment was directed at just three commodities— corn, soybeans, and cotton (Traxler et al., 2005, personal communication). These are the largest U.S. crop seed markets, where large private returns encourage the most private R&D. Public R&D is spread more evenly across the full set of crops, including crops such as wheat where social returns to research justify public investment but high development costs and low appropriability reduce private returns. Public R&D is also more geographically dispersed, with experiment stations in each State. Because agriculture is heavily influenced by factors that vary across regions such as climate, natural resource quality, and transportation costs, the geographic dispersion of public R&D addresses challenges in a broader set of agricultural production systems.

Public R&D focuses more heavily on basic or "pre-technology" research that is difficult to patent and commercialize, while most private R&D emphasizes applied research and product development. For example, decades of publicly funded research in molecular genetics and biotechnology in the latter half of the 20th century enabled private firms to develop new techniques with commercial potential in agriculture.

Complemented by new policies that strengthened intellectual property rights over crop varieties and genetic traits, these scientific and technological advances led to a surge in private-sector investment in crop breeding and biotechnology R&D (Fuglie et al., 1996). Between 1980 and 2010, R&D spending by U.S. seed and crop biotechnology companies rose from $100 million per year to more than $2 billion annually in constant 2010 dollars (Fuglie et al., 2011).

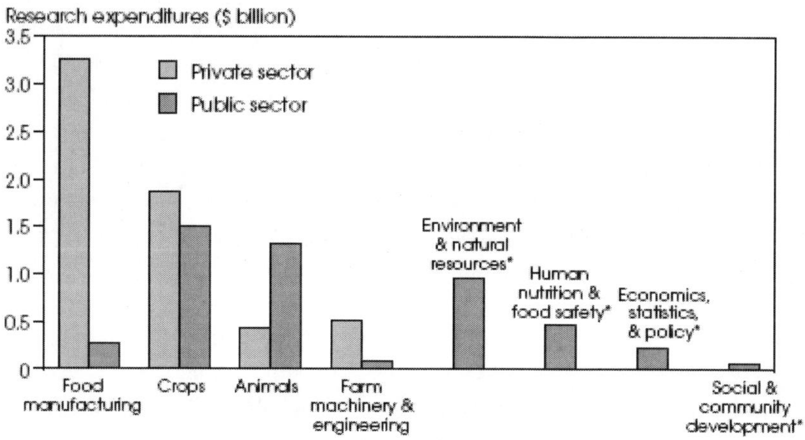

*Private-sector investment is likely to be greater than zero, but reliable figures could not be obtained.

Source: Public R&D is from the U.S. Department of Agriculture (2007) based on Knowledge Area classifications. Private R&D figures are from Fuglie et al. (2011).

Figure 3. Public research addressed a broader set of research topics than private research in 2006.

A comprehensive survey of public and private plant breeding in 1996 found that most private crop breeding R&D focused on downstream development of cultivars (cultivated crop varieties) while public R&D tended toward long- and medium-term research projects on plant breeding methods, identification of new traits, and germplasm enhancement to improve parent breeding materials (fig. 4). As the private sector increased investment in applied breeding, public research shifted to more basic R&D activities (Fuglie and Walker, 2001). Other studies of plant breeding (Pardey et al., 1996; Lesser and Kolady, 2011) emphasize the contributions of genetic material from publicly maintained elite germplasm used to develop private cultivars.

Other studies that have looked at public and private R&D throughout the U.S. agricultural research system also found their relationship to be complementary rather than duplicative. Both Wang et al. (2009) and Tokgoz (2006) found that private investments in agricultural R&D were significantly correlated with past trends in public spending on agricultural and life sciences research. The authors of these studies attributed this correlation to (1) scientific advances from public R&D that created new technological opportunities for private firms to commercialize, and (2) the higher education and training function of university research that expanded the pool of scientific human capital for private firms to hire. A third study, using citations data from scientific publications, found that total citations from private research to university publications increased, and that firms citing more university research had a higher likelihood of producing patentable technologies (Toole and King, 2011).

The agricultural chemicals industry also demonstrates the complementarity of private and public R&D. Companies that develop new agricultural chemical products devote most of their R&D toward regulatory testing, product registration, and product marketing, and many producers of existing "generic" products do minimal R&D. Public R&D in crop protection often addresses basic research questions about the biology of crops and crop pests and applied research on the use of chemicals in production systems, development of resistance, and integrated pest management practices (Fuglie et al., 2011).

Animal agriculture has not seen a similar surge in private R&D investment, but public and private roles are nonetheless complementary. The public sector performs a majority of animal breeding and genetics R&D, while most of the private spending on animal agriculture shown in Figure 3 targets pharmaceutical animal health products (Fuglie et al., 2011). Private R&D in breeding and genetics is lower in animal agriculture relative to crops in part because of incentives that vary markedly among animal commodities (higher for poultry, lower for large ruminants) and consumer acceptance and regulatory frameworks for genetically modified animals that remain unsettled.

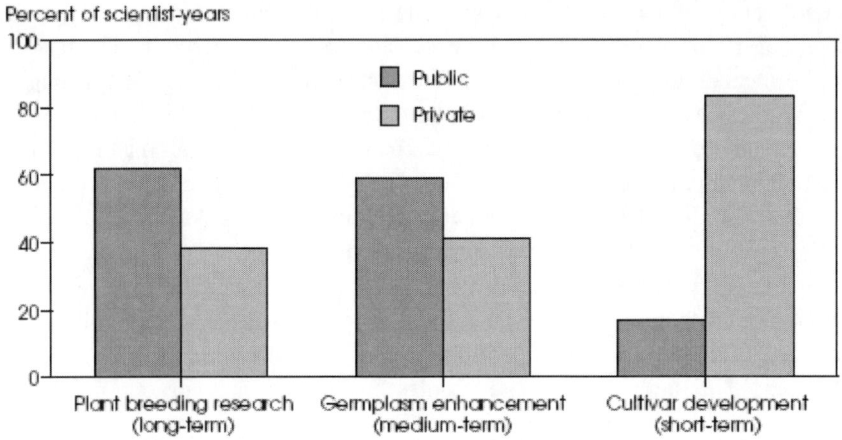

Source: Frey (2000). Estimates are based on a survey of 90 Federal and State research institutions and 329 private seed companies. For each research area, the shares sum to 100 percent of the scientists reported in the survey to be working in that area.

Figure 4. In crop breeding, the public sector emphasized longer term research and the private sector dominated development of cultivated varieties in 1996.

Policies and Partnerships Combine Strengths to Enhance the Agricultural Research System

With a shift of overall food and agricultural R&D toward the private sector, public research institutions have developed new strategies for partnering with private firms. New laws and policies provide vehicles for formal collaboration between public and private research institutions. Technology transfer policies encourage interaction of public and private R&D at different phases of technological development. When researchers at publicly funded institutions produce discoveries with commercial potential, they can obtain patents on these discoveries and license them to companies interested in further commercial development. This allows public researchers to focus on their public research missions and hand off commercialization to private technology partners. Exclusive licenses of patented technology reduce appropriation risks and raise private returns for commercial partners, thereby increasing incentives for firms to invest in downstream development, marketing, and production.

When the nature of R&D requires closer interaction, Cooperative Research and Development Agreements (CRADAs) provide a contractual mechanism for public and private researchers to collaborate. The use of both patent licensing and CRADAs by the USDA has grown since these arrangements were first enabled through legislation in the 1980s (fig. 5).Through these agreements, government-owned and operated labs can partner with private firms to share data, facilities, and research results (Guston, 1998; Day-Rubenstein and Fuglie, 2000). Firms engaged in a CRADA are able to protect proprietary information and negotiate exclusive licenses for CRADA inventions.

While the use of CRADAs, patenting, and other technology transfer instruments could potentially divert public research away from its central mission (National Research Council, 1995), such arrangements may also draw private capital into areas that serve important societal needs but where market failures are most evident, like human nutrition and natural resource conservation (Day-Rubenstein and Fuglie, 2000).

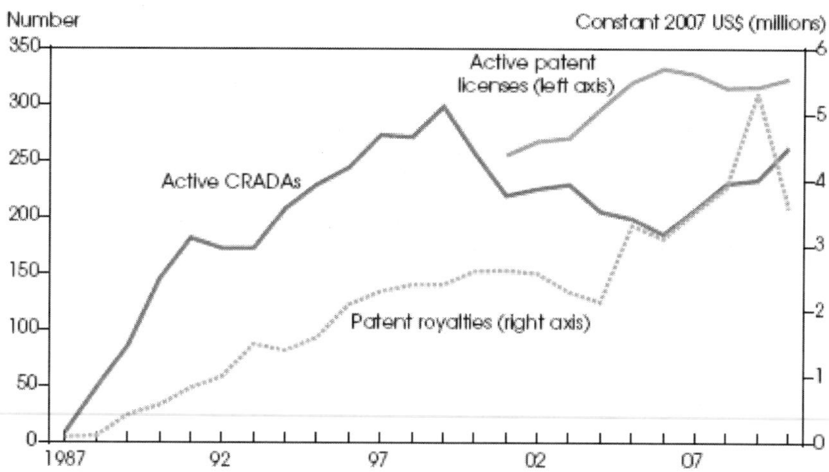

Source: Data from 2001 are from Annual Reporting on Technology Transfer in USDA, Office of Technology Transfer, Agricultural Research Service. Earlier data are from Heisey et al. (2006).

Figure 5. USDA uses public-private R&D partnerships to facilitate technology transfer.

Another mode of public-private cooperation is when private interest groups, such as farmer commodity organizations, levy a small assessment on their sales to support research at public institutions. Federal legislation has permitted agricultural producer groups to make such "checkoffs" mandatory for research and promotion, but support for research through this mechanism in the United States has been minimal (Alston and Pardey, 1996). Some countries, notably Australia, have experimented with a model in which the government matches producer levies for public research. Recently, Australia has also instituted a system of end-point royalties based on sales of improved varieties as a way of enforcing intellectual property rights and encouraging more private investment in crop development. In the Australian experience, these institutional innovations appear to have strengthened private support for agricultural research and improved its efficiency and coordination (Alston, Gray, and Bolek, 2012).

References

Adams, James D., and Roger Clemmons. 2008. "The Origins of Industrial Scientific Discoveries." NBER Working Paper 13823. National Bureau of Economic Research, Cambridge, MA.

Alston, J., and P. Pardey. 1996. *Making Science Pay: The Economics of Agricultural R&D Policy*. Washington, DC: American Enterprise Institute.

Alston, J., M. Andersen, J. James, and P. Pardey. 2010. Persistence Pays: US Agricultural Productivity Growth and the Benefits from Public R&D Spending. New York: Springer.

Alston, J., R. Gray, and K. Bolek. 2012. "Farmer-Funded R&D: Institutional Innovations for Enhancing Agricultural Research Investments." Working Paper, Department of Agricultural and Resource Economics, University of California at Davis.

Day-Rubenstein, Kelly, and Keith Fuglie. 2000. "The CRADA Model for Public-Private Research and Technology Transfer in Agriculture," *Public-Private Collaboration in Agricultural Research* (K. Fuglie and D. Schimmelpfennig, eds.). Ames: Iowa State University Press, pp. 155-174.

Frey, Kenneth J. 2000. "A National Strategy for Plant Breeding in the United States," *Public-Private Collaboration in Agricultural Research* (K. Fuglie and D. Schimmelpfennig, eds.). Ames: Iowa State University Press, pp. 77-98.

Fuglie, Keith O., Nicole Ballenger, Kelly Day, Cassandra Klotz, Michael Ollinger, John M. Reilly, Utpal Vasavada, and Jet Yee. 1996. *Agricultural Research and Development: Public and Private Investments Under Alternative Markets and Institutions.* Agricultural Economics Report 735, U.S. Department of Agriculture, Economic Research Service, May.

Fuglie, Keith O., and David Schimmelpfennig, eds. 2000. *Public-Private Collaboration in Agricultural Research: New Institutions and Economic Implications.* Ames: Iowa State University Press.

Fuglie, Keith O., and Thomas S. Walker. 2001. "Economic Incentives and Resource Allocation in U.S. Public and Private Plant Breeding," *Journal of Agricultural and Applied Economics* 33(3):459-473.

Fuglie, Keith O., James M. MacDonald, and V. Eldon Ball. 2007a. *Productivity Growth in U.S. Agriculture.* Economic Brief No. 9, U.S. Department of Agriculture, Economic Research Service, Sept.

Fuglie, Keith O., and Paul W. Heisey. 2007b. *Economic Returns to Public Agricultural Research.* Economic Brief No. 10. U.S. Department of Agriculture, Economic Research Service, Sept.

Fuglie, Keith O., Paul Heisey, John King, Carl E. Pray, Kelly Day-Rubenstein, David Schimmelpfennig, Sun Ling Wang, and Rupa Karmarkar-Deshmukh. 2011. *Research Investments and Market Structure in the Food Processing, Agriculture Input and Biofuel Industries Worldwide.* Economic Research Report 130. U.S. Department of Agriculture, Economic Research Service.

Guston, David H. 1998. "Technology Transfer and the Use of CRADAs at the National Institutes of Health," *Investing in Innovation* (L.M. Branscomb and J.H. Keller, eds.). Cambridge: The MIT Press, pp. 221-249.

Heisey, Paul W., John L. King, Kelly Day-Rubenstein, and Robbin Shoemaker. 2006. *Government Patenting and Technology Transfer.* Economic Research Report 15. U.S. Department of Agriculture, Economic Research Service.

Heisey, Paul, Sun Ling Wang, and Keith O. Fuglie. 2011. *Public Agricultural Research Spending and Future U.S. Agricultural Productivity Growth: Scenarios for 2010-2050.* Economic Brief No. 17, U.S. Department of Agriculture, Economic Research Service, July.

Lesser, William, and Deepthi Elizabeth Kolady. 2011. "Disease Resistance of Wheat Varieties: Can Private Varieties Withstand the Pressure?" Economics Research International, vol. 2011, Article ID 575192.

National Research Council. 1995. Allocating Federal Funds for Science and Technology. Washington, DC: National Academies Press.

National Science Foundation. 2009. *National Patterns of R&D Resources.* Division of Science Resources Statistics, National Science Foundation, Washington, DC.

Pardey, P., J. Alston, J. Christian, and S. Fan. 1996. "Summary of a Productive Partnership: The Benefits from U.S. Participation in the CGIAR." International Food Policy Research Institute, Environment and Production Technology Division Discussion Paper.

Tokgoz, Simla. 2006. "Private Agricultural R&D in the United States," *Journal of Agricultural and Resource Economics* 31(2):212-238.

Traxler, Greg, A. Acquaye, K. Frey, and A.M. Thro. 2005. Public Sector Plant Breeding Resources in the US: Study Results for the Year 2001. U.S. Department of Agriculture, National Institute for Food and Agriculture.

Toole, Andrew, and John King. 2011. "Industry-Science Connections in Agriculture: Do Public Science Collaborations and Knowledge Flows Contribute to Firm-Level Agricultural Research Productivity?" Selected paper, Agricultural and Applied Economics Association Annual Meeting, July 24-26, Pittsburgh, PA.

U.S. Department of Agriculture, National Institute of Food and Agriculture. 2007. *Inventory of Agricultural Research.* Current Research Information System.

U.S. Department of Agriculture, Economic Research Service. 2012. "Agricultural Research Funding in the Public and Private Sectors." http://www.ers.usda.gov/data-products/agricultural-research-fundingin-the-public-and-private-sectors.aspx

Wang, C., Y. Xia, and S.T. Buccola. 2009. "Public Investment and Industry Incentives in Life Science Research," *American Journal of Agricultural Economics* 91(2):374-388.

In: U.S. Agricultural Preparedness ...
Editor: Warren C. Young
ISBN: 978-1-62618-437-4

Chapter 4

PUBLIC AGRICULTURAL RESEARCH SPENDING AND FUTURE U.S. AGRICULTURAL PRODUCTIVITY GROWTH: SCENARIOS FOR 2010-2050*

U.S. Department of Agriculture

MAJOR FINDINGS

- By 2050, global agricultural demand is projected to grow by 70-100 percent due to population growth, energy demands, and higher incomes in developing countries. Meeting this demand from existing agricultural resources will require raising global agricultural total factor productivity (TFP)[1] by a similar level. Maintaining the U.S. contribution to global food supply would also require a similar rise in U.S. agricultural TFP.
- TFP growth in U.S. agriculture is predicated on long-term investments in public agricultural research and development (R&D). Productivity growth also springs from agricultural extension, farmer education, rural infrastructure, private agricultural R&D, and technology

* This is an edited, reformatted and augmented version of U.S. Department of Agriculture, Economic Research Service, Economic Brifef Number 17, dated July 2011.

transfers, but the force of these factors is compounded by public agricultural research.

- The rate of TFP growth (and therefore output growth) of U.S. agriculture has averaged about 1.5 percent annually over the past 50 years. Stagnant (inflation-adjusted) funding for public agricultural research since the 1980s may be causing agricultural TFP growth to slow down, although statistical analyses of productivity growth trends are inconclusive.
- ERS simulations indicate that if U.S. public agricultural R&D spending remains constant (in nominal terms) until 2050, the annual rate of agricultural TFP growth will fall to under 0.75 percent and U.S. agricultural output will increase by only 40 percent by 2050. Under this scenario, raising output beyond this level would require bringing more land, labor, capital, materials, and other resources into production.
- Additional public agricultural R&D spending would raise U.S. agricultural productivity and output growth. Raising R&D spending by 3.73 percent annually (offsetting the historical rate of inflation in research costs) would increase U.S. agricultural output by 73 percent by 2050. Raising R&D spending by 4.73 percent per year (1-percent annual growth in inflation-adjusted spending) would increase output by 83 percent by 2050.

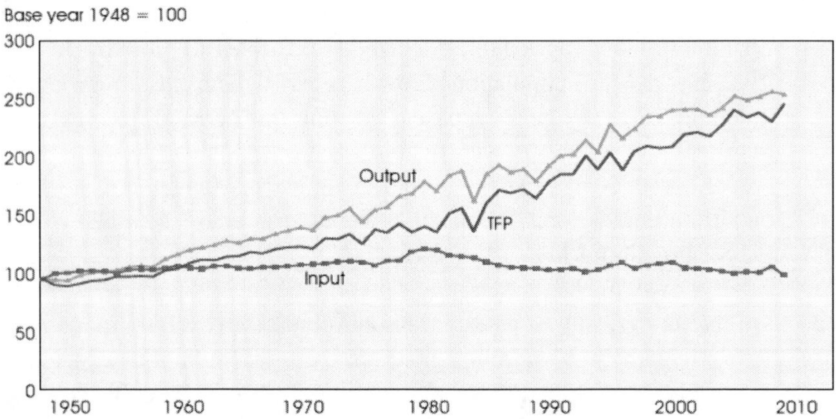

Source: U.S. Department of Agriculture, Economic Research Service.

Figure 1. U.S. agricultural output, input, and TFP indexes.

PRODUCTIVITY GROWTH ACCOUNTS FOR NEARLY ALL U.S. AGRICULTURAL GROWTH

Total factor productivity (TFP) is a good indicator of technological change. It measures the efficiency with which all inputs (land, labor, capital, and materials) are combined to produce total outputs (all crop and livestock commodities). In U.S. agriculture, growth in TFP is nearly synonymous with growth in output since the overall size of the resource base has barely increased over the last 60 years. Between 1948 and 2008, the average annual growth rate in U.S. agricultural output was 1.58 percent and the average annual growth rate in TFP was 1.52 percent. As a result, total agricultural output in 2008 was 2.5 times that of 1948 (figure 1).

R&D SPENDING SPURS TFP GROWTH

Growth in TFP is strongly associated with the adoption of new technologies that raise yields or lower costs. Public investment in agricultural R&D is a major source of new agricultural technology; it also complements (raises returns to) other productivity-enhancing activities like extension, education, infrastructure, and private R&D. While private-sector investment in agricultural R&D also contributes new technology, much of this contribution is already accounted for in the way the inputs are measured in constructing the estimates of TFP.[2] Moreover, studies have shown that private R&D depends critically on government and university investments in science and technology.[3] Thus, even if the quality-adjusted measures of TFP do not fully net out the contributions of private R&D, if private R&D follows (responds positively to) trends in public R&D, then simulations using public R&D as the primary policy lever for future productivity growth should not be unduly biased. Nonetheless, the simulations reported here depend on these assumptions about private R&D: that its impacts are partially netted out of the TFP measure and that it follows trends in public R&D.

Statistical models of TFP growth treat R&D spending as "knowledge capital." As with physical capital, an R&D investment can affect productivity growth for several years (even decades), but eventually depreciates as technologies grow ineffective (e.g., new pests and diseases evolve) or obsolete. R&D investments typically begin boosting TFP within 3-5 years, with benefits peaking after 10 to 20 years, and with some impacts lasting as long as 50 years (Huffman and Evenson, 2006; Alston et al., 2010).

AGRICULTURAL R&D SPENDING SCENARIOS ACCOUNT FOR INFLATION IN RESEARCH COSTS

Between 1927 and 2009 (the period for which we have data), total spending by Federal and State agricultural research institutes on productivity-related agricultural research (excluding R&D on postharvest, environmental, and rural development issues) rose at an average annual rate of 7.49 percent in nominal terms. If adjusted by the Consumer Price Index, then research spending rose by 4.37 percent per year. However, if inflation adjustments reflect the cost of doing research (i.e., scientist salaries, laboratory equipment, etc.), then real research spending rose by only 3.03 percent per year (i.e., research costs have been rising faster than the average price level of goods and services in the economy).

Estimates of R&D knowledge capital are based on real research spending (adjusted by the rate of inflation in the cost of research resources). Growth in public agricultural research spending slowed dramatically after the early 1980s and has declined by more than 20 percent since peaking in 1994 (figure 2). The growth in the stock of agricultural knowledge capital, which follows trends in research spending with a lag, peaked around 2005. Since then, knowledge capital stock has started to decline as current spending (creation of new knowledge capital) has failed to keep up with knowledge capital depreciation.

We simulate three possible scenarios for future public agricultural R&D expenditure for 2010-2050 (see Box for a description of the simulation model). These scenarios include total expenditures on productivity-oriented agricultural research by USDA research agencies and State agricultural experiment and veterinary medicine colleges from all sources of funding (USDA, other Federal and State governments, and others).

Scenario 1: Public research spending is constant in nominal terms at the 2005-09 average ($2.5 billion per year). This implies a decline in real research expenditure by 3.73 percent per year (the inflation rate in the cost of research resources over 1983-2009).[4] In other words, under this scenario, inflation reduces the effective amount of research effort.

Scenario 2: Public research spending is constant in real terms at the 2005-09 average. This implies that nominal expenditures for public agricultural research rise by about 3.73 percent per year to offset inflation in the cost of research.

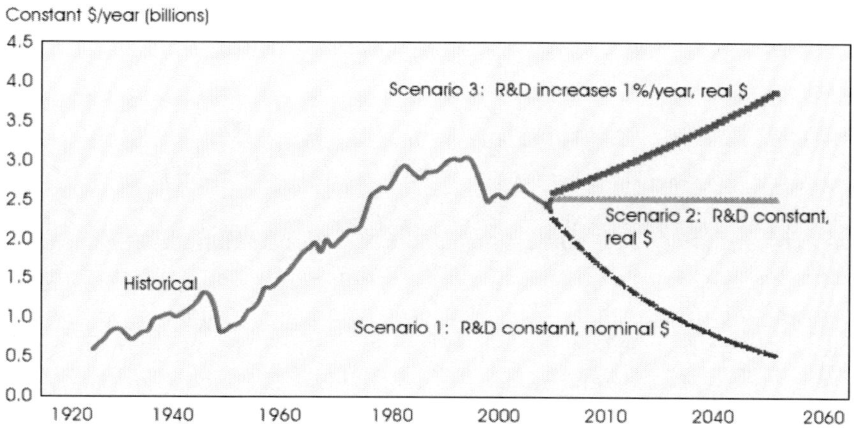

Source: Historical data from Huffman (2010) and 2010-2050 data from U.S. Department of Agriculture, Economic Research Service simulations.

Figure 2. Productivity-oriented agricultural research expenditures.

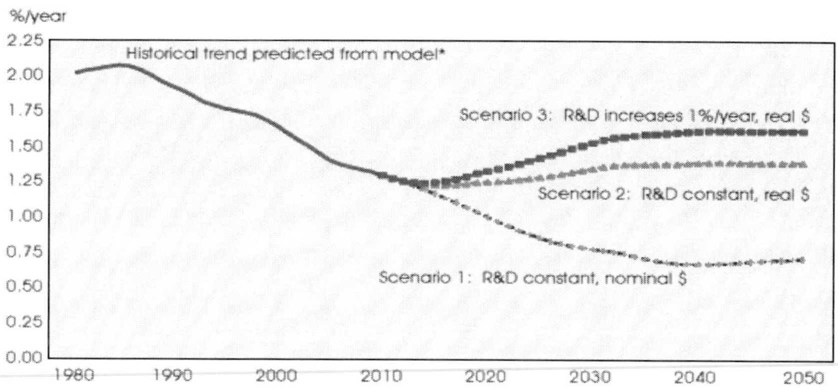

* The graph shows the TFP growth rate predicted from the statistical model rather than the actual TFP growth rate for 1980-2008. Actual TFP growth is highly variable, ranging between -15% and + 15% for individual years. This variability is mainly due to weather.

Figure 3. TFP growth rate projections.

Scenario 3: Public research spending increases by 1 percent per year in real terms (implying a nominal increase of about 4.73 percent per year). This 1-percent real growth rate is roughly equivalent to the rate of U.S. population growth (a proxy for growth in domestic agricultural demand). Real research expenditures rise, but more slowly than the rapid growth in real spending observed before the 1980s.

R&D SPENDING MUST RISE TO MAINTAIN HISTORICAL AGRICULTURAL TFP GROWTH RATE

The resulting TFP projections under the three public R&D spending scenarios are shown in figures 3 and 4. Figure 3 shows the projected *annual growth rate* in agricultural TFP. The estimated historical trend based on our model shows a declining rate of TFP growth since peaking at just over 2 percent per year in the mid-1980s. By 2008, annual TFP growth was about 1.3 percent.[5] In each of the scenarios, rates of TFP growth increase or decrease from present levels and then level off after 2030 or 2040. Under scenario 1, the TFP growth rate declines to under 0.75 percent per year. Under scenario 2, TFP growth stabilizes at about 1.4 percent per year. Under scenario 3, TFP growth accelerates and reaches 1.6 percent per year by 2040.

Figure 4 translates TFP growth into an agricultural TFP index (base year of 2008=100) under alternative funding scenarios. It also shows the agricultural output level (relative to 2008 production) achievable if agriculture's current resource base remains unchanged. Under scenario 3, by 2050 U.S. agricultural output would be 83 percent higher than 2008 levels and under scenario 2, output would grow by 73 percent. As a point of comparison, current estimates suggest global food demand will increase between 70 and 100 percent by 2050 (Food and Agriculture Organization, 2006). Under scenario 1, however, with constant R&D funding (in nominal terms), U.S. agricultural output would increase by only about 43 percent between 2008 and 2050 using the current agricultural resource base. This would barely outpace the projected growth in U.S. population. By contrast, U.S. agricultural output would likely keep up with the growth in global demand under the higher R&D investment scenarios.

Another recent study (Alston et al. 2010) used different data and modeling assumptions to independently derive statistical relationships between public agricultural R&D spending and TFP growth. Their findings are broadly similar to our own: public spending on agricultural R&D will need to rise to maintain historical rates of productivity growth in U.S. agriculture.

Implications of a decline in the long-term growth rate in U.S. agricultural productivity are significant for both the United States and the world. Slower productivity growth will likely cause agricultural prices to rise, which would reduce global economic welfare and raise poverty in urban areas and among food-deficit rural households, especially in developing countries (Nelson et al., 2010). Rising prices would also increase pressure to expand agricultural cropland at home and abroad and input use on cropland would likely intensify. Such agricultural resource expansion and intensification could come with significant environmental costs, such as further impairment of soil and water quality, loss of biodiversity, and increased emissions of greenhouse gases (Foresight, 2011). Finally, agricultural exports may also decline as U.S. farmers lose competitiveness in international markets (Ball et al., 2010).

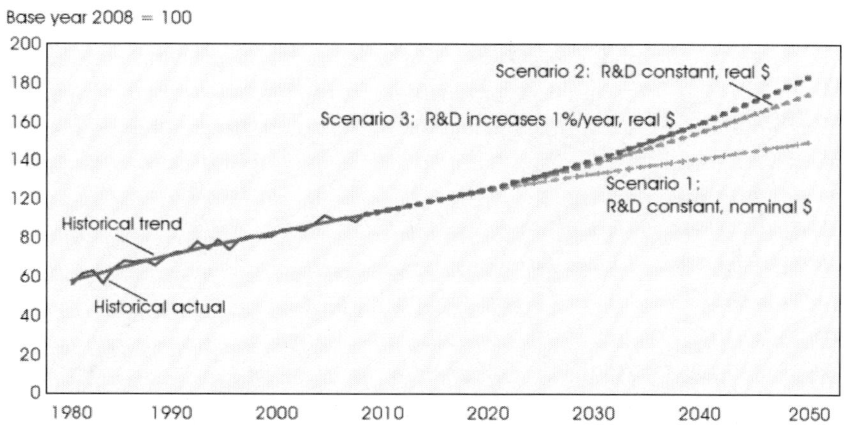

Figure 4. TFP index projections.

HOW THE ESTIMATES WERE DERIVED

We use a model developed by ERS (Wang et al., 2010) along with research expenditures data from Wallace Huffman at Iowa State University. Huffman isolates the component of total public agricultural R&D associated with raising crop and livestock productivity and excludes other kinds of R&D (Huffman et al., 2001; Huffman and Evenson, 2006; Huffman, 2010).

Statistical models that link TFP to these investments treat R&D spending as "knowledge capital." Like physical capital, knowledge capital contributes to growth over a long period of time but eventually depreciates; it requires new R&D investment to be maintained and expanded. We use the assumptions developed by Huffman and colleagues to convert annual R&D investments, or flows, into R&D capital, or stocks. Under these assumptions, R&D investments influence R&D stocks, and therefore TFP, for up to 35 years.

We assume alternative scenarios for future public R&D funding to simulate future growth in U.S. agricultural total factor productivity to 2050. We use statistical relationships based on different productivity and investment patterns across States over 1980-2004 to estimate the effects of R&D stocks on productivity growth. These estimated relationships are used to project future TFP growth patterns given alternative assumptions about future R&D spending. Changes in current levels of public R&D spending affect future TFP growth only gradually since most of today's "knowledge capital stock" is the result of past, accumulated investment in R&D. In order to model the effects of different scenarios for R&D spending on future TFP growth, we hold the contributions to agricultural productivity from other sources—such as agricultural extension, farmer education, infrastructure, economies of scale, and technology transfer—constant and allow only the effects from public agricultural R&D to change.

REFERENCES

Alston, Julian M., Matthew A. Andersen, Jennifer S. James, and Philip G. Pardey. 2010. Persistence Pays: U.S. Agricultural Productivity Growth and the Benefits from Public R&D Spending. Springer.

Ball, V. Eldon, Jean-Pierre Butault, Carlos San Juan, Ricardo Mora. 2010. "Productivity and international competitiveness of agriculture in the European Union and the United States." Agricultural Economics 41: 611-627.

Economic Research Service. "Data Set: Agricultural Productivity in the United States," Agricultural Research and Productivity Briefing Room. Available at http://www.ers.usda.gov/Data/AgProductivity/ (accessed June 1, 2011).

Food and Agriculture Organization. 2006. World agriculture: towards 2030/2050, Interim report. Food and Agriculture Organization of the United Nations, Rome, June. http://www.fao.org/docrep/ oo9/a0607e/ a0607e00.htm

Foresight. 2011. The Future of Food and Farming: Challenges and Choices for Global Sustainability. Final Project Report. The Government Office for Science, London, United Kingdom.

Fuglie, Keith O. and Thomas S. Walker. 2001. "Economic Incentives and Resource Allocation in U.S. Public and Private Plant Breeding. Journal of Agricultural and Applied Economics 33 (3): 459-473.

Huffman, W. E., A. McCunn, and J. Xu. 2001. "Public Agricultural Research Expenditures with an Agricultural Productivity Emphasis: Data for 48 States, 1927-1995." Iowa State University, Department of Economics, Staff Paper.

Huffman, W. E., and R.E. Evenson. 2006. Science for Agriculture—A Long-Term Perspective. Second edition, Blackwell Publishing.

Huffman, W.E. 2010. "Measuring Public Agricultural Research Capital and Its Impact on State Agricultural Productivity in the U.S." Iowa State University, Department of Economics, Mimeo. (Updates data from Huffman et al., 2001)

Nelson, Gerald C., Mark W. Rosegrant, Amanda Palazzo, Ian Gray, Christina Ingersoll, Richard Robertson, Simla Tokgoz, Tingju Zhu, Timothy B. Sulser, Claudia Ringler, Siwa Msangi and Liangzhi You. 2010. Food Security, Farming and Climate Change to 2050. International Food Policy Research Institute, Washington, DC.

Wang, Chenggang, Yin Xia, and Steven Buccola. 2009. "Public Investment and Industry Incentives in Life-Science Research." American Journal of Agricultural Economics 91(2): 374-388.

Wang, Sun Ling, Eldon Ball, Lilyan Fulginiti, and Alejandro Plastina. 2010. "Benefits of Public Research Spill-in, Extension, and Road Networks in U.S. Agriculture." (Under revise and resubmit with a peer reviewed journal).

End Notes

[1] Total factor productivity (TFP) is the broadest measure of productivity. It compares the total output of a sector to the total land, labor, capital, and material inputs used to produce that output. Increases in TFP imply more output is forthcoming from a given level of inputs, or, equivalently, fewer inputs are required to produce the same output. Growth in TFP is considered to be an indicator of the rate of technical change in a sector.

[2] New technologies from private R&D are partially embodied in the quality of the inputs (chemicals, machinery, seed, etc.) purchased by farmers. The TFP measure used in this

study adjusts for such quality improvements in these inputs. In other words, it already nets out many of the productivity effects of private R&D.

[3] Public-sector research opens up new technological opportunities for the private sector to commercialize. One recent study found that "government expenditures in both basic biological research and agricultural and medical science create substantial spillovers for private firms...Indeed, opportunities created through public research are the principal source of growth in industry life sciences" (Wang, Xia and Buccola, 2009). Fuglie and Walker (2001) also found complementarity between public and private investments in crop genetic improvement, with public research in basic plant breeding spurring more private investment in crop variety development.

[4] While the average rate of inflation in research costs over 1927-2009 was 4.46 percent, this has slowed in more recent years. To project research cost inflation in the future, we chose the average inflation rate in research costs during 1983-2009, or 3.73 percent per year.

[5] This apparent decline in TFP growth is based on the model used to project future TFP growth. All other factors were kept fixed except for the changes in R&D "knowledge capital" stock; therefore, the decline in the TFP growth rate reflects the erosion of R&D stocks, as hastened by the stagnation in R&D funding that began in the 1980s (see figure 2). Thus, the slowing of TFP growth in figure 3 after the mid-1980s reflects this model assumption; actual agricultural TFP growth may not have slowed if other contributing factors offset the decline in R&D capital.

In: U.S. Agricultural Preparedness ... ISBN: 978-1-62618-437-4
Editor: Warren C. Young

Chapter 5

Economic Returns to Public Agricultural Research*

U.S. Department of Agriculture

Introduction

Over the last several decades, the U.S. agricultural sector has sustained impressive productivity growth. The Nation's agricultural research system, including Federal-State public research as well as private-sector research, has been a key driver of this growth. Economic analysis finds strong and consistent evidence that investment in agricultural research has yielded high returns per dollar spent.

These returns include benefits not only to the farm sector but also to the food industry and consumers in the form of more abundant commodities at lower prices.

While studies using different methods and coverage give a range of estimates of returns to agricultural research, there is a consensus that the payoff from the government's investment in agricultural research has been high.

---

* This is an edited, reformatted and augmented version of a United States Department of Agriculture, Economic Research Service publication, Economic Brief Number 10, dated September 2007.

Agricultural Productivity Growth Is a Driving Force in U.S. Agriculture

Whether measured as crop yield per acre, milk and meat yield per animal, or average output per farm worker, the productivity of U.S. agriculture is among the highest in the world. The reasons for gains in productivity are many. For example, corn yields have increased through greater use of agricultural inputs, such as more fertilizers and machinery per acre of land. The development of new technology, however, also has helped boost yields. New technologies not only made inputs more effective but allowed them to be combined in new and better ways. For example, precision agriculture allows farmers to target fertilizer use more judiciously, raising crop yields and reducing costs. ERS has developed statistical series to distinguish the contribution of changes in input use from that of other factors affecting the growth of the agricultural sector. In particular, ERS has developed an index measure of Total Factor Productivity (TFP) to distinguish the effects of innovation and related factors on the growth of agricultural output.[1] In the long run, growth in TFP is the primary source of new wealth creation in the economy. Therefore, trends in agricultural TFP may provide an indication of the longrun performance of the sector.

For the period 1948-2004, the index value for output reached 266 in 2004, compared with the base year, 1948, meaning that total agricultural production in 2004 was 2.66 times higher than it was in 1948 (fig. 1). Over the same period, aggregate use of inputs (land, labor, capital, and intermediate inputs) in agriculture actually decreased slightly. Although the use of some inputs, such as fertilizer and machinery, increased, these increases were offset by reductions in cropland and, especially, the amount of labor employed in agriculture. Overall, the amount of crop and livestock output produced per unit of (aggregate) input, or the change in TFP, increased dramatically. From 1948 to 2004, agricultural productivity growth was strong in each decade, allowing output to grow with little or no increase in inputs throughout the period.

Today's Investment in Research Drives Tomorrow's Growth in Productivity

Analysts have found a strong link between investments in research and innovation and agricultural productivity growth. However, there is a long lead time between the research stage of a new technology and the point at which that technology is adopted and begins to affect productivity. Therefore, tracking research investments in food and agriculture is a good indicator of the likely future trends in agricultural productivity growth. In the United States, both the public sector (Federal and State governments) and the private sector invest heavily in agricultural research. From 1971 to the late 1990s, public and private agricultural research spending combined rose from just over 3 percent to about 7 percent of agricultural GDP (analysts have not been able to systematically track private-sector investments in food and agricultural research since 1998 due to unavailability of comparable data)(fig. 2). In constant 2001 dollars, annual spending on food and agricultural research in the United States increased from about $4.6 billion to $8.8 billion between 1970 and the late 1990s, with the private sector accounting for most of this growth. Public spending for agricultural research was mostly flat (after adjusting for inflation) between 1978 and 1998 but showed some renewed growth during 1998-2004.

Note that public spending on agricultural research includes not only support for productivity-oriented research but also research on natural resources, food nutrition and safety, rural development, and economics. About 60 percent of public agricultural research is for enhancing productivity and the rest is for other objectives or fields of study. About 70 percent of private research in the late 1990s was oriented to farm production and about 30 percent went to food manufacturing.

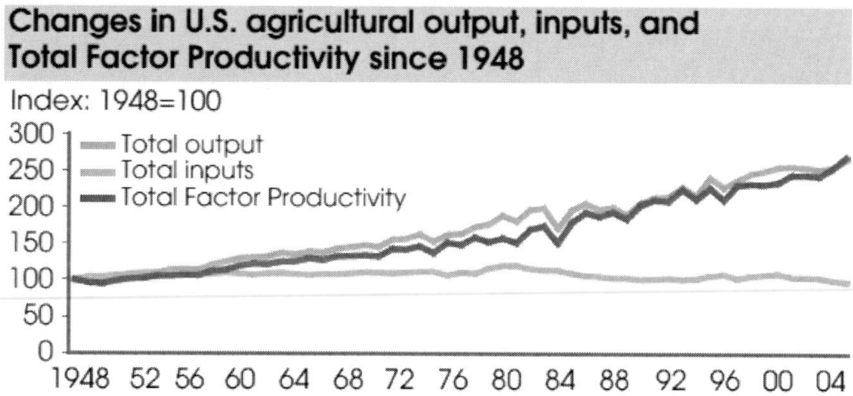

Source: USDA/ERS.

Figure 1. Changes in U.S. agricultural output, inputs, and Total Factor Productivity since 1948.

Economic Studies Find High Social Returns to Investments in Agricultural Research

Economic assessments of payoffs from public investments in agricultural research have attempted to determine the "social rate of return" to this expenditure (see box, "How Economists Evaluate Returns to Agricultural Research"). This is reported as a percent return on each dollar spent on research. The return is "social" because it includes all of the economywide benefits from higher productivity. These returns benefit not only farmers but also the food industry and consumers, who gain from more abundant and lower cost commodities. As a benchmark, social returns to public expenditures are often compared with the return to U.S. Treasury Bonds as a measure of the opportunity cost of public funds. Historically, the real return on U.S. government securities (the real return equals the nominal return minus the inflation rate) has been around 3-4 percent per year.

Against this benchmark, economic studies find that public expenditures on agricultural research have yielded real returns several magnitudes higher. Some of these studies have estimated returns to research on particular commodities or in particular States, but several have assessed returns to investment in the Federal-State public agricultural research system as a whole, for various periods of the 20th century. For 35 studies published over 1965-2005 that were reviewed by Professors Wallace Huffman (Iowa State University) and Robert Evenson (Yale University), the median estimate of the social rate of return was 45 percent per year (table 1). As a rough approximation, this implies that each dollar spent on agricultural research returned about $10 worth of benefits to the economy.[2] Although these studies have produced a range of estimates of the rate of return to agricultural research (estimates are sensitive to methods and assumptions), they all agree that the return to public agricultural research has been significantly higher than the benchmark return from government securities.

ERS analyzed findings from 27 studies that assessed the rate of return to public agricultural research in the United States over various periods of the 20th century. Estimated rates of return varied depending on study methodology and coverage, but the major share ranged from 20 to 60 percent (fig. 3). Specific results from several of these studies (those published in peer-reviewed journals or as Ph.D. theses) are reported in table 2. Most of the studies were published in the 1970s through the 1990s, which suggests the estimates are becoming somewhat dated. We have found only one study estimating returns to agricultural research published since 2000.

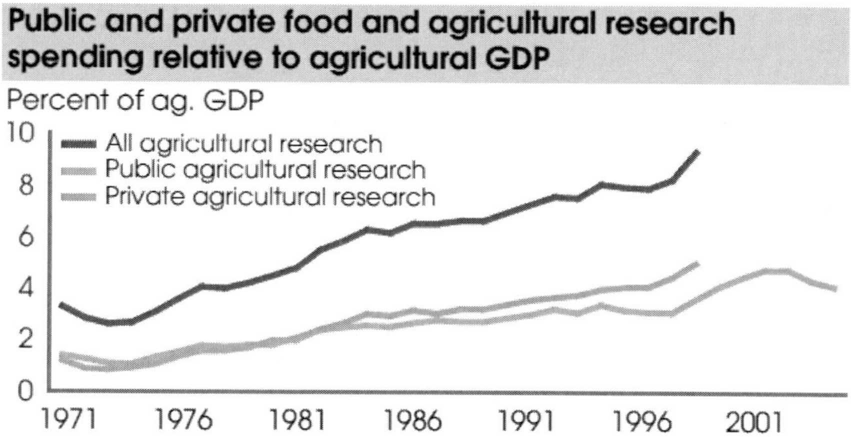

Source: USDA, ERS (public and private agricultural research expenditures) and Economic Report of the President (agricultural GDP calculated as a 3-year moving average).

Figure 2. Public and private food and agricultural research spending relative to agricultural GDP.

Several other conclusions can be drawn from the literature on returns to agricultural research:

- *Returns to research have been high for most crop and livestock commodities.* Economic studies that have estimated returns to research on particular crop or livestock commodities have usually found evidence of high returns, although returns do vary by commodity and over time. However, there is little clear evidence that research on certain commodities has given consistently higher returns than research on other commodities.

How Economists Evaluate Returns to Agricultural Research

Economic evaluations of agricultural research are based on comparisons between (i) public and private investments in agricultural knowledge creation and dissemination, and (ii) long-term changes in agricultural productivity. Conceptualization of this process is as follows:

- Expenditures on agricultural research generate new knowledge that eventually leads to improved technology that is adopted by farmers.
- Technology adoption increases average productivity (the output of crop and livestock commodities per unit of land, labor, capital, and intermediate inputs employed in production).
- Higher productivity of agricultural resources leads to lower costs, higher production, and/or exit of some resources (such as labor) from the agricultural sector.
- Given physiological limits to per capita demand for food, higher agricultural production leads to lower commodity prices, passing some of the technology-induced cost reductions on to the food industry and consumers. Thus, benefits of productivity-enhancing agricultural research are shared between the farm and nonfarm sectors of the economy.

The accompanying figure illustrates the typical time pattern of development, adoption, and eventual obsolescence of agricultural technology. As shown, a public (or private) institution invests in the development of a new technology (such as a new crop variety with disease resistance) and spends several years working on that effort ("research costs" in the figure). After about 7 years, the technology is successfully developed and farmers begin to adopt it. Costs are still incurred in extension efforts, and benefits grow as more farmers adopt the technology and reap higher yields or lower production costs. In the figure, it takes about 8 years (from year 7 until year 15) for the technology to be fully adopted and benefits maximized thereafter. But after some time, the technology eventually goes out of use, either because something better replaces it or because it loses its effectiveness (due to buildup of resistance in the pathogen, for the research endeavor weighs the size of the research and extension costs against the economic benefits from technology adoption, discounting the benefit and cost streams to measure them in terms of their "present value.

There are two main approaches used to estimate economic returns to agricultural research:

Statistical analysis relating past expenditures on research to current changes in productivity. These models try to establish a statistical correlation between when, where, and what research was done and productivity gains in agriculture. The analysis is usually done at a fairly aggregate level and covers a long period of time. For example, a study may look at the annual pattern of changes in agricultural productivity at the national or State level over several decades and then relate this in a regression model to investments in public and private agricultural research and extension over a similar period (with research investments beginning earlier than productivity changes to account for the lag time between research and technology adoption). These studies also examine effects of other factors that may contribute to productivity growth, like investments in rural education, extension, and infrastructure. If regression analysis finds positive and significant correlations between research expenditures (appropriately lagged) and productivity changes, then this is taken as evidence of a causal relationship. An estimate of the rate of return to research is derived from the regression coefficients.

Project evaluation methods tracing the development and dissemination of innovations. An early example of this approach was a study by Zvi Griliches (Dept. of Econ., University of Chicago) in the 1950s on the returns to research on hybrid maize. He estimated the benefits of hybrid maize by measuring the economic value of higher maize yield made possible from this innovation. On the cost side, he estimated the cost of research and extension (by both the public and private sectors) beginning with the work of George Schull of the Carnegie Institution and Donald Jones of the Connecticut Agricultural Experiment Station, who developed the theory of hybrid vigor and invented the double-cross method of hybrid seed production.

The case study method provides a clearer cause-and-effect relationship between agricultural research and productivity growth than the regression method. But the case study method has largely been limited to analysis of research "success stories."

Regression methods, on the other hand, assess the system at a more aggregate level and take into account expenditures on research that may or may not lead to successes and, therefore, tend to give a more balanced measure of average returns to a research system.

Both approaches involve estimating relationships between the size of investment in research and the economic value of increased productivity, taking into account the appropriate time dimension between when research is done and when economic benefits are realized, such as the case depicted in the figure. Estimates of social returns to research may be overstated if undesirable outputs (e.g., environmental degradation) are not taken into account. Similarly, social returns may be understated if new technology reduces undesirable outputs.

Flows of research costs and benefits over time

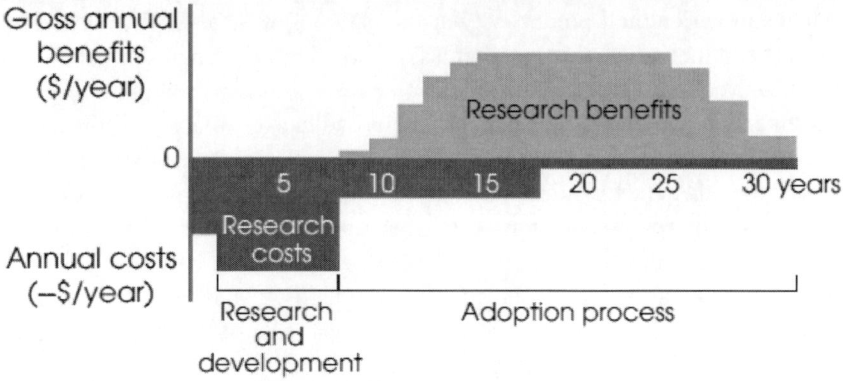

Source: Alston, Norton, and Pardey, 1995.

Some of the most challenging aspects of these models are:

- *Lags:* Identifying the appropriate lag relationship between when research is done and when productivity growth occurs.
- *Spillovers:* Accounting for knowledge or research "spillovers" across geographic space. Spillovers occur when research done in one State, region, or country contributes to new knowledge or technology that is used in another geographic area.
- *Attribution:* Accounting for the many elements that come together to contribute to the development and application of new technology to agriculture. In addition to publicly funded agricultural research, contributions include those made by basic sciences, innovations from the private sector, farmer education, the training role of extension services, and improvements to rural infrastructure. These institutional sources are often complementary, and failure to account for the contribution of one source may overattribute observed gains in productivity to another source. Including all these sources in a model may give an indication of the relative importance of each source (and the relative rate of return to each). Some studies go even further to try to distinguish returns to agricultural research done by Federal or State institutions, or even by different Federal funding instruments (e.g., formula versus competitive grants). But putting finer and finer distinctions among sources of innovation and types of research expenditure places a heavy burden on the data.

- There appear to be significant social returns to private agricultural research.Manufacturing industries, especially for chemical, machinery, pharmaceuticals, and food and feed processing (and, more recently, for biotechnology), are significant sources of new technology for agriculture. The private sector earns a return from its investment in agricultural research in the profit it makes on sales of improved chemicals, machines, feeds, breeds, and seeds to farmers. If the private sector could capture all of the benefits of its technology through the price it charges for its products, then private research would not register as a source of improvement in farm productivity (the value of higher output would be offset by the higher cost of inputs). Only a few studies have assessed social returns to private agricultural research, and findings suggest that private research does contribute significantly to measured productivity growth in agriculture (table 1), implying that the private sector is able to capture only a share of the productivity benefits from its technology.
- *Agricultural research generates long-term benefits.*That an investment in research entails a long lag time before it produces tangible economic impact is well understood. It is also clear that the lag is longer for more basic research than for more applied research, and that a sizeable share of research undertaken may never be applied to technology development that is adopted by farmers. Economists have used regression models to try to determine the average lag of agricultural research undertaken by the public sector in the United States and to determine how long this research continues to contribute to productivity growth until it becomes obsolete. As more data have been accumulated on agricultural research and growth, it has become possible to estimate more sophisticated models on the statistical relationship between these variables. Current research on this topic suggests that, on average, public agricultural research undertaken today will begin to noticeably influence agricultural productivity in as little as 2 years and that its impact could be felt for as long as 30 years.
- Agricultural knowledge or research "spillovers" across State and national boundaries are significant.Although much research in agriculture is oriented toward the ecological conditions of a particular State or region, there is strong evidence (from both statistical models and observation of trade in agricultural inputs that embody new technology) that agricultural research done in one location affects

productivity in other regions or even other countries. Spillovers from livestock research are generally greater than spillovers from crop research because livestock production is less constrained by agro-ecological factors like soils and climate.

Table 1.Summary estimates of the rate of return to U.S. agricultural research

Item	Studies, 1965-2005	Mean estimate	Median estimate
Social rate of returns to public agricultural research	35	53	45
Social rate of returns to private agricultural research	4	45	45

Source: USDA, ERS, using data from Huffman and Evenson, 2006, and Fuglie et al.,1996.

Table 2. Estimates of the rate of return to Federal-State investment in agricultural research

Studies on the aggregate crop-animal sector

Study	Authors	Pub. year	Publication	Period	Coverage	ROR estimate		
						Mid	Low	High
1	Huffman & Evenson	2006	Am J Ag Econ	1970-1999	Crops & animals	56	49	62
2	Gopinath & Roe	2000	Econ Innov & Tech	1960-1991	Crops & animals	37		
3	Makki et al.	1999	J Policy Modeling	1930-1990	Crops & animals	27		
4	White	1995	J Ag Appl Econ	1950-1991	Crops & animals	40		
5	Chavas & Cox	1992	Am J Ag Econ	1950-1982	Crops & animals	28		
6	Norton & Ortiz	1992	J Production Agric	1987, state-level comp.	Crops & animals	58		
7	Yee	1992	J Ag Econ Res	1931-1985	Crops & animals	54	49	58
8	Braha & Tweeten	1986	Tech. Bull., Ok SU	1959-1982	Crops & animals	47		
9	Lyu, White, Liu	1984	S J Ag Econ	1949-1981	Crops & animals	66		
10	White & Havlicek	1982	Am J Ag Econ	1943-1977	Crops & animals	22	7	36
11	Davis	1979	PhD thesis,	1949-1959	Crops &	83	66	100

Study	Authors	Pub. year	Publication	Period	Coverage	ROR estimate		
						Mid	Low	High
			UMN		animals			
11	Davis	1979	PhD thesis, UMN	1964-1974	Crops & animals	37		
12	Knutson & Tweeten	1979	Am J Ag Econ	1949-1972	Crops & animals	38	28	47
13	Lu et al.	1979	Tech. Bull., ERS	1939-1972	Crops & animals	27	23	30
14	Bredahl & Peterson	1976	Am J Ag Econ	1937-1942	Crops & animals	56		
14	Bredahl & Peterson	1976	Am J Ag Econ	1947-1957	Crops & animals	51		
14	Bredahl & Peterson	1976	Am J Ag Econ	1957-1962	Crops & animals	49		
14	Bredahl & Peterson	1976	Am J Ag Econ	1967-1972	Crops & animals	34		
15	Cline	1975	PhD thesis, Ok SU	1939-1948	Crops & animals	46	41	50
16	Evenson	1968	PhD thesis, U Chic	1949-1959	Crops & animals	47		
17	Peterson	1967	J Farm Econ	1915-1960	Crops & animals	23	21	25
18	Griliches	1964	Amer Econ Rev	1949-1959	Crops & animals	33	25	40

Studies on components of the agricultural sector

Study	Authors	Pub. year	Publication	Period		ROR estimate		
						Mid	Low	High
6	Norton & Ortiz	1992	J Production Agric	1987, state-level comp.	Beef & swine	55		
6	Norton & Ortiz	1992	J Production Agric	1987, state-level comp.	Dairy	95		
6	Norton & Ortiz	1992	J Production Agric	1987, state-level comp.	Poultry	46		
6	Norton & Ortiz	1992	J Production Agric	1987, state-level comp.	Grain crops	31		
6	Norton & Ortiz	1992	J Production Agric	1987, state-level comp.	Potatoes, cotton &			
					tobacco	34		
6	Norton & Ortiz	1992	J Production Agric	1987, state-level comp.	Vegetables &			
					melons	19		
6	Norton & Ortiz	1992	J Production Agric	1987, state-level comp.	Fruits & nuts	33		
23	Haygreen et al.	1986	Forest Prod J	1972-1981	Forest products	25	14	36

Table 2. (Continued)

Study	Authors	Pub. year	Publication	Period		ROR estimate		
						Mid	Low	High
22	Bengston	1984	Forest Science	1975, state-level comp.	Forest products	21	19	22
19	Smith et al.	1983	J NE Ag Econ	1978, state-level comp.	Beef & swine	22		
19	Smith et al.	1983	J NE Ag Econ	1978, state-level comp.	Dairy	25		
19	Smith et al.	1983	J NE Ag Econ	1978, state-level comp.	Poultry	61		
21	Schmitz & Seckler	1970	Am J Ag Econ	1958-1969	Tomato harvester	42	37	46
17	Peterson	1967	J Farm Econ	1915-1960	Poultry	23	21	25
20	Griliches	1958	J. Poli. Econ	1940-1955	Corn	38	35	40
20	Griliches	1958	J. Poli. Econ	1940-1957	Sorghum	20		

ROR = Rate of return. Most studies assessed productivity change over a period of years. “State-level comp.” refers to studies that compared productivity among States at a point in time against past research expenditures in those States. Contact the authors for a complete set of references.

Source: USDA, ERS, using data from Huffman and Evenson, 2006, and Fuglie et al.,1996.

- There are conjectures but little empirical evidence that returns to agricultural research have fallen over time. Over time, it may become increasingly difficult for scientists and engineers to find new ways of raising crop yield or reducing the need for farm labor. In addition, as agricultural productivity reaches increasingly high levels, it is probably necessary to devote more resources to simply maintaining these levels rather than achieving even greater productivity. The need for more maintenance research does not imply that returns to research would diminish (since preventing productivity from falling has economic value) but it would imply that more research dollars would be required to prevent that rate of productivity growth from falling. So far, economic studies have not found any clear indication that the long-term growth rate in agricultural productivity has declined.
- There has been little empirical work assessing returns to agricultural research on nonmarket objectives (natural resource quality, food safety, economics, and policy). The agricultural research system devotes considerable resources to research on natural resources, food nutrition and safety, economics and statistics, and other objectives not

directly related to raising farm productivity. While economic studies have documented specific impacts of some of this research, these nonmarket outcomes have not been incorporated into estimates of returns to research.

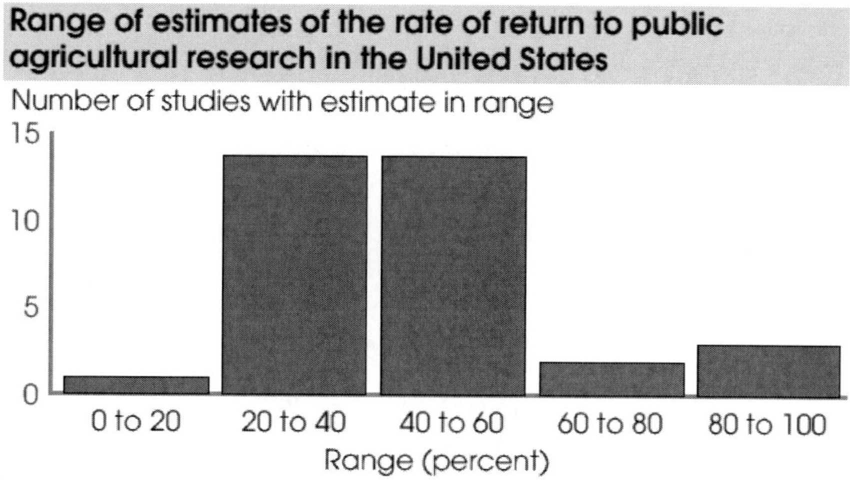

Source: USDA, ERS, using data from Huffman and Evenson, 2006.

Figure 3. Range of estimates of the rate of return to public agricultural research in the United States.

LONG-TERM PROSPECTS FOR U.S. AGRICULTURAL PRODUCTIVITY GROWTH

One key factor that will shape future growth in U.S. agricultural productivity is the Nation's investment in agricultural research, both in amount and in how well this research investment is used.

The U.S. agricultural research system consists of Federal, State, and private-sector elements. Policies that shape this system include not only government spending for agricultural research but also policies that encourage private-sector research and farm adoption of new technology. A measure of the effectiveness of these policies is the level of public and private research expenditures relative to agricultural Gross Domestic Product (GDP). Agricultural GDP measures the value added by the agricultural sector. Combined public and private spending on agricultural research as a share of agricultural GDP more than doubled between 1970 and 1998 to just over 7 percent (see fig. 2). Comparable data for private-sector research are not available for the years since 1998, but public spending on agricultural research as a percentage of agricultural GDP continued to grow until 2002 and then fell in 2003 and 2004. So long as public and private investment in agricultural research continues to keep pace with or grow faster than agricultural GDP, and so long as this investment continues to earn the high social rates of return as in the past, prospects for the agricultural sector to maintain productivity growth are favorable.

This Brief Is Drawn from ...

Alston, Julian M., George W. Norton, and Philip G. Pardey. Science Under Scarcity: Principles and Practices for Agricultural Research Evaluation and Priority Setting. Ithaca, NY: Cornell University Press, 1995.

Fuglie, Keith O., Nicole Ballenger, Kelly Day, Cassandra Klotz, Michael Ollinger, John Reilly, Utpal Vasavada, and Jet Yee. Agricultural Research and Development: Public and Private Investments Under Alternative Markets and Institutions. Agricultural Economics Report 735, U.S. Department of Agriculture, Economic Research Service, May 1996, www.ers.usda. gov/publications/aer735/

Griliches, Zvi. "Research Costs and Social Returns: Hybrid Corn and Related Innovations," Journal of Political Economy 66 (1958): 419-31.

Huffman, Wallace E., and Robert E. Evenson. Science for Agriculture: A Long-Term Perspective, 2nd Edition, Ames, IA: Blackwell Publishing, 2006.

U.S. Department of Agriculture, Economic Research Service. Agricultural Research and Productivity Briefing Room, ERS website, www.ers.usda. gov/briefing/agresearch/

End Notes

[1] For more information on trends in agricultural productivity growth, see the ERS Economic Brief Productivity Growth in U.S. Agriculture.

[2] An internal rate of return can be converted (approximately) into a benefit-cost ratio by dividing the social rate of return by the opportunity cost of capital. If we allow 4 percent (the longrun real yield of U.S. Government securities) to represent the cost of social capital, then a rate of return to research of 40 percent would imply a benefit-cost ratio of about 10 to 1.

INDEX

D

E

F

G

H

M

N

O

P

Q

R

S

T

U

V

W